CELLULAR NEURAL NETWORKS AND ANALOG VLSI

edited by

Leon O. Chua
University of California, Berkeley

Glenn Gulak
University of Toronto

Edmund Pierzchala
Analogy, Inc.

Angel Rodríguez-Vázquez
Universidad de Sevilla

A Special Issue of
ANALOG INTEGRATED CIRCUITS AND SIGNAL PROCESSING
An International Journal
Volume 15, No. 3 (1998)

KLUWER ACADEMIC PUBLISHERS
Boston / Dordrecht / London

ANALOG INTEGRATED CIRCUITS AND SIGNAL PROCESSING

An International Journal

Volume 15, No. 3, March 1998

Special Issue: Cellular Neural Networks and Analog VLSI

Guest Editors: Leon O. Chua, Glenn Gulak, Edmund Pierzchall and Angel Rodriguez-Vázquez

Distributors for North America:
Kluwer Academic Publishers
101 Philip Drive
Assinippi Park
Norwell, Massachusetts 02061 USA

Distributors for all other countries:
Kluwer Academic Publishers Group
Distribution Centre
Post Office Box 322
3300 AH Dordrecht, THE NETHERLANDS

Library of Congress Cataloging-in-Publication Data

Cellular neural networks and analog VLSI / edited by Leon O. Chua ...
[et al.].
p. cm.
"A special issue of Analog integrated circuits and signal processing, an international journal, volume 15, no. 3 (1998)."
Includes bibliographical references.
ISBN 0-7923-8125-4 (alk. paper)
1. Neural networks (Computer science) 2. Integrated circuits--Very large scale integration. I. Chua, Leon O., 1936- II. Analog integrated circuits and signal processing. Special issue.
QA76.87.C44 1998
621.39'9--dc21 97-53185
CIP

Printed on acid-free paper.

Printed in the United States of America

Analog Integrated Circuits and Signal Processing, 15, 225–226 (1998)

Editorial

Hardly anyone would question the importance of digital systems in information procesing. The "digital" revolution is both a result and a cause ot technological changes, especially in semiconductor manufacturing. A few wonder if, or when, analog systems will become extinct. So long as "digital" is merely a way of coding continuous information, a convenient abstraction which allows the ignoring of intermediate states, the question whether "analog" is more important than "digital" or vice-versa, is irrelevant.

In a great many situations it is *convenient* or *beneficial* to ignore intermediate states in a system. It is simplistic, to say the least, to assume that the only interesting or practically useful perspective on information processing, is one that originates that way. Stating it differently, there is no justification for the belief, not unheard of today, that only digital information processing has a bright future.

The present issue contains some recent results of research in a broad class of analog systems called "Cellular Neural Networks" (CNNs), which consist of an *array* of *locally-interconnected* analog circuit units, called "cells."

Although the CNN paradigm allows the cells to be of arbitrary complexity, the current generation of CNN chips uses only simple first-order cells (one capacitor per cell) in order to maximize the cell density (array size). Although the 8-bit accuracy of current analog CNN chips is adequate for most CNN-based image processing applications, much higher accuracy for specialized applications can be achieved via digital emulations of the CNN's nonlinear differential equations, at the cost of a reduction in speed and cell density.

CNNs, introduced in 1987 by Prof. Leon O. Chua and his collaborators, have sparked worldwide interest and have become a very fruitful and active area of research. The current intense activity[1] on CNN research was triggered by the recent generalization of the CNN paradigm by Prof. T. Roska and Prof. L. Chua, called the CNN Universal Machine, which is a fully programmable von Neumann stored program supercomputer capable of executing tera (10^{12}) XPS (analog instructions per second) on a single chip. This invention has resulted in a dramatic three-orders-of-magnitude improvement in *speed*, *power*, and *area* over the equivalent conventional DSP implementations.

In this issue, CNN analog-circuit implementations are of primary interest.

A companion issue of Field-Programmable Analog Arrays (FPAAs) is in preparation. CNNs (at least their analog implementations) and FPAAs share the "analog" nature, though both address rather different applications. Also, some FPAAs have cellular architectures, which can be used to *implement* CNNs, but can also be utilized for entirely different applications.

Leon O. Chua
Edmund Pierzchala
Glenn Gulak
Angel Rodríguez-Vázquez

Notes

1. See the Proceedings of the bi-annual IEEE International Workshop on Cellular Neural Networks and their Applications (held in 1990, 1992, 1994, and 1996), as well as several special issues on Cellular Neural Networks of the IEEE Transactions on Circuits and Systems (March 1993) and the International Journal of Circuit Theory and Applications (1992 and 1996). See also: L. O. Chua "CNN: A Paradigm for Complexity," *Int. J. of Bifurration and Chaos*, Vol. 7, no. 9, September 1997.

Leon Chua is currently a Professor of electrical engineering and computer sciences at the University of California, Berkeley. His research interests are in the areas of general *nonlinear network and system theory*. He has been a consultant to various electronic industries in the areas of nonlinear network analysis, modeling, and computer aided design. He is the author of *Introduction to Nonlinear Network Theory* (McGraw-Hill, 1969), and a coauthor of the books *Computer*

Aided Analysis of Electronic Circuits: Algorithms and Computational Techniques (Prentice-Hall, 1975), *Linear and Nonlinear Circuits* (McGraw-Hill, 1987), and *Practical Numerical Algorithms for Chaotic Systems* (Springer-Verlag, 1989). He has published many papers in the area of nonlinear networks and systems. He served as editor of the IEEE *Transactions on Circuits and Systems* from 1973 to 1975 and as President of the *IEEE Society on Circuits and Systems* in 1976. He is presently the editor of the International Journal of Bifurcation and Chaos and a deputy editor of the *International Journal of Circuit Theory and Applications*.

Professor Chua is the holder of six U.S. patents. He is also the recipient of several awards and prizes, including the 1967 IEEE Browder J. Thompson Memorial Prize Award, the 1973 IEEE W.R.G. Baker Prize Award, the 1974 Frederick Emmons Terman Award, the 1976 Miller Research Professorship, the 1982 Senior Visiting Fellowship at Cambridge University, the 1982/83 Alexander von Humbolt Senior U.S. Scientist Award at the Technical University of Munich, the 1983/84 Visiting U.S. Scientist Award at Waseda University, Tokyo, the IEEE Centennial Medal in 1985, the 1985 Myril B. Reed Best Paper Prize, the 1985 and 1989 IEEE Guillemin-Cauer Prizes, and the 1995 M.E. Van Valkenburg Award. In 1986 he was awarded a *Professeur Invité* International Award at the University of Paris-Sud from the French Ministry of Education. He was also awarded a *Doctor Honoris Causa* from the Ecole Polytechnique Fédérale-Lausanne, Switzerland, in 1983, an Honorary Doctorate from the University of 1992, a *Doctor Honoris Causa* from the Technical University of Budapest, Hungary, in 1994, a *Doctor Honoris Causa* from the University of Santiago de Compostela, Spain, in 1995, a *Doctor Honoris Causa* from the University of Frankfurt, Germany, in 1996, and a *Doctor Honoris Causa* from the Technical University of Iasa, Romania, in 1997.

bio and photo not available at time of print

Edmund Pierzchala

Glenn Gulak is a professor in the Department of Electrical and Computer Engineering at the University of Toronto. He is a senior member of the IEEE and a registered professional engineer in the province of Ontario. His research interests are in the areas of circuits, algorithms and VLSI architectures for digital communications and signal processing applications. He has received several teaching awards for undergraduate courses taught in both the Department of Computer Science and the Department of Electrical in both the Department of Computer Science and the Department of Electrical and Computer Engineering at the University of Toronto. Dr. Gulak received his Ph.D. from the University of Manitoba while holding a Natural Sciences and Engineering Research Council of Canada Postgraduate Scholarship. From 1985 to 1988 he was a research associate in the Information Systems Laboratory and the Computer Systems Laboratory at Stanford University. He has served on the ISSCC Signal Processing Technical Subcommittee since 1990 and currently serves as Program Secretary for ISSCC.

bio and photo not available at time of print

Angel Rodríguez-Vázquez

Analog Integrated Circuits and Signal Processing, 15, 227–237 (1998)

A 16 × 16 Cellular Neural Network Universal Chip: The First Complete Single-Chip Dynamic Computer Array with Distributed Memory and with Gray-Scale Input-Output

J. M. CRUZ[1] AND L. O. CHUA[2]

[1] *Department of Electrical Engineering and Computer Sciences, University of California, Berkeley; Current address: Sun Microsystems, Inc.*
[2] *Department of Electrical Engineering and Computer Sciences, University of California, Berkeley*

Abstract. This paper presents a 16 × 16 Cellular Neural Network Universal Chip with analog input and output ports, which can read in and process gray-scale images in the analog domain. The chip contains about 5,000 analog multipliers and has been fabricated in a 0.8 μm CMOS process.

Key Words: CNN, CNN universal chip, computer array, CMOS, analog, nonlinear dynamics

1. Introduction

The Cellular Neural Network (CNN) architecture was presented in [1] and [2]. This architecture allows parallel analog processing of images using an array of locally interconnected cells with fixed weights. The local interconnection feature allows efficient VLSI implementations [3][4], and many CNN chips have been reported since the first operational CNN chip was presented in 1991 [5].

In 1993 the CNN architecture was augmented, incorporating programmable weights, local storage, and local logic. The new augmented architecture is called Cellular Neural Network Universal Machine [6], as it has proven to be able to solve all the algorithms that can be executed by a Turing Machine [7].

Different types of chip implementations of the CNN Universal Machine or parts of it have been reported recently [8]–[12], and they are usually referred to as CNN Universal Chips. These implementations include: (a) designs which operate according to the original CNN continuous-time analog-amplitude dynamical equations given in [1], as in the design reported in [12] and in a programmable chip design without logic memory reported in [11]; (b) a design which operates according to a modified continuous-time analog-amplitude CNN equation in which the state variable is clipped to unity and the final chip output is binary [8]; (c) a design which emulates the CNN dynamics by using discrete time samples but maintaining analog-amplitude [9]; and (d) a digital implementation which emulates the original CNN dynamics by using discrete time samples and discrete amplitude values [10]. These implementations vary in their area and speed efficiency, with reported data showing higher efficiency in both respects for circuits implementing the CNN dynamics directly in the analog domain. These reported chips also vary in other aspects of their functionality, including their capability to locally store and logically manipulate binary images, and their capability to input, store, process, and output not only binary images but gray-scale images.

In this paper we present the first CNN Universal Chip with the capability to input, store, process, and output gray-scale images in the analog domain combined with the capability to locally store and logically process binary images. The chip has been fabricated in a 0.8um 3-metal 1-poly technology of HP.

Each cell implements the dynamics of the CNN equation with a state resistance by state capacitance product below 90 ns. For typical applications the dynamics of the entire array settles to a constant state in 200 to 250 ns. Each cell has a computing power, excluding input-output operations, of 4 to 5 million pixels per second and the entire chip has a peak computing capability of 1 billion pixels per second. The maximum electrical power consumption of the entire chip when operating at 5 Volts is 0.3 Watts.

2. Chip Architecture

Figure 1 shows a photograph of the chip, packaged in a 132 PGA. The main components of the chip are as follows.

Fig. 1. Photograph of packaged chip.

- Array of 16×16 CNN cells for simultaneous processing of image blocks of 14×14 pixels.
- Analog input port bus with 14 lines to load external binary or gray-scale images column by column.
- Serial analog output port to read out gray-scale processed images or to trace in real time the transient evolution of the output variable of any selected cell during the analog processing.
- Digital output port bus with 14 lines to fast read out binary processed images.
- Analog control port with 19 analog lines to set the analog CNN template coefficients.
- Digital control port to set the logic operations and to start and stop analog process.

Figure 2 shows a diagram of the internal placement of the main components in the chip. In the central part of the chip is the 16×16 array of cells, which is divided in an inner array of 14×14 *processing* cells and in 60 *boundary* cells located at the outer edges and corners. This structure can be used to perform parallel processing of blocks 14×14 pixels at a time with different boundary conditions.

At the left side of the cell array is a 14-line analog input bus for the external loading of analog data into the two internal analog memories of each one of the processing cells. The loading of these memories is done column by column. At the bottom of the chip diagram is the column selector decoder, which is used to indicate which column is being selected for input or output. At the right side of the array are: (1) a set of 14 logic buffers which drive a 14-line logic parallel output and (2) a 14:1 analog MUX connected to an analog buffer providing an analog serial output.

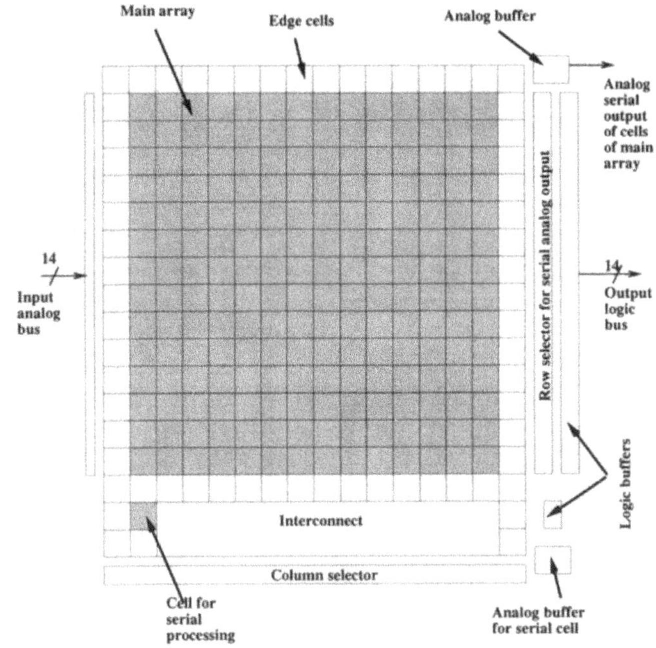

Fig. 2. Chip structure.

Finally, at the lower left end of the chip is an additional cell configured for serial processing of images. The boundary conditions for this cell are set by eight boundary cells, three of them are at the lower left corner of the chip, two at the lower right corner of the chip, and the other three are shared with three of the bottom boundary cells of the main array. This serial processing cell has its own input and output ports.

3. Description of Processing Cells

The core element of the chip is the CNN cell. Figure 3 shows a schematic of the implemented cell which contains capacitors, resistors, and variable gain VCCSs (multipliers). Observe that the cell provides eight output currents to the eight neighboring cells and receives eight input currents from the eight neighboring cells.

Each cell $c(i, j)$ has three analog variables: the input variable, the state variable and the output variable. The indexes i and j denote the absolute row and column position of the cell in the array. Each cell $c(i, j)$ at row i and column j provide output currents to a set of neighboring cells $c(i+k, j+l)$ located at row $i+k$ and column $j+l$ of the array, where the indexes k and l denote the relative row and column position of the neighboring cells to respect $c(i, j)$. In this chip each cell provides outputs to its eight closest neighboring cells located at the relative positions given by $(k, l) = (-1, -1)$, $(-1, 0)$, $(-1, +1)$, $(0, -1)$, $(0, +1)$, $(+1, -1)$, $(+1, 0)$, $(+1, +1)$. Likewise, each cell $c(i, j)$ receives eight input currents, one for each of its eight neighboring cells.

The cell dynamics are determined by a set of 19 weights, called template elements, which control the gain of 18 VCCSs and the value of an independent current source. These template elements are divided into nine feedback template elements, nine feedforward elements and an independent term. The nine feedback elements, denoted by $a(k, l)$ where both k and l vary from -1 to $+1$, determine the gain of nine VCCSs controlled by the cell state voltage $x(i, j)$ which provides a current into each one of the nine cells in the local 3×3 neighborhood of iteration (the cell itself plus the eight neighboring cells). The nine feedforward elements, denoted by $b(k, l)$ where both k and l vary from -1 to $+1$, determine the gain of nine VCCSs controlled by

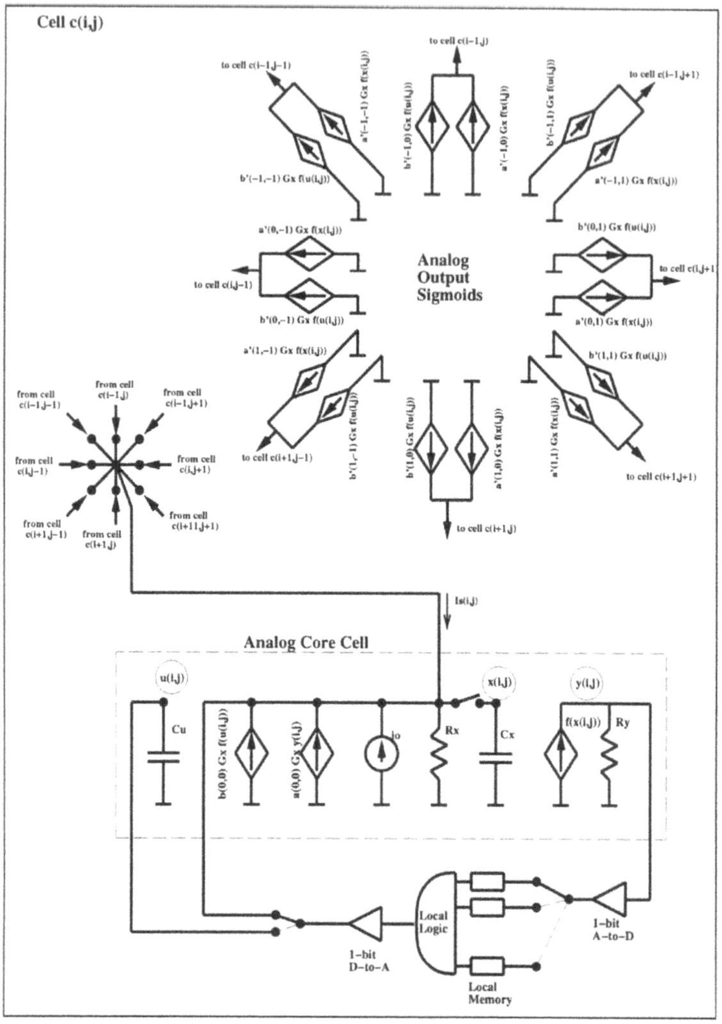

Fig. 3. Ideal circuit diagram of one cell.

the cell input voltage $u(i, j)$ which provide a current into each one of the nine cells in the local 3×3 neighborhood of iteration. The independent term, denoted by i_o, represents the value of an independent source providing an offset current into the state node of the cell.

Figure 4 shows a more detailed diagram of the cell. The state variable $x(i, j)$ drives a differential output nonlinear amplifier with a saturation-type transfer characteristic $f(.)$ [1]. The amplifier output drives nine four-quadrant multipliers, that perform the multiplication with the nine feedback template coefficients $a(k, l)$. Likewise the input variable $u(i, j)$ drives another amplifier, which drive another set of nine multipliers that perform the multiplication with the nine feedforward template coefficients $b(k, l)$. The outputs of the two set of multipliers are added by pairs, and the results are provided in current mode into the state nodes of the nine cells in the local 3×3 neighborhood.

Figure 5 shows a schematic of two differential pairs

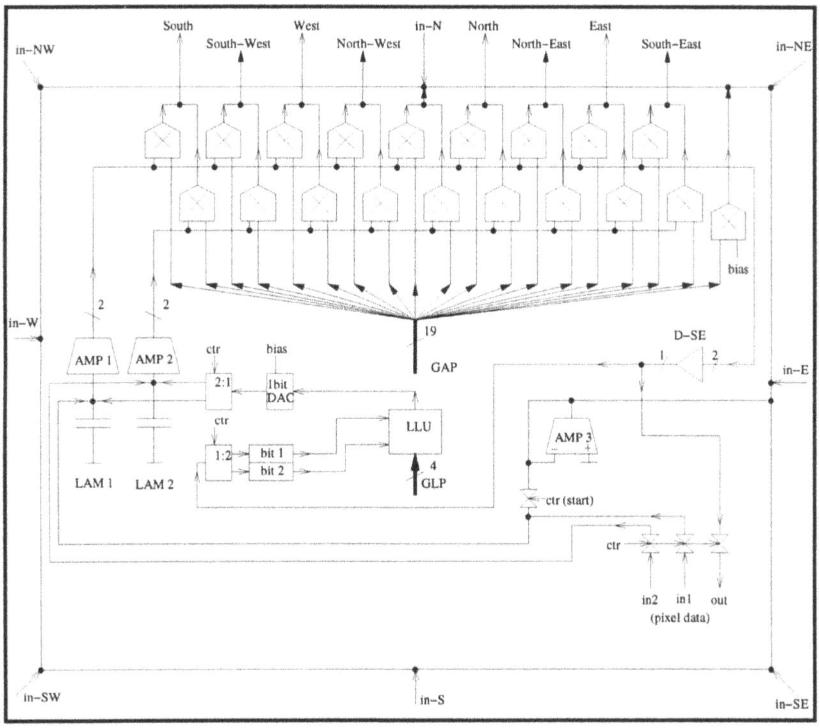

Fig. 4. Cell structure.

that are used as amplifiers connected to a set of two multipliers. The bias current and the transistor sizes of the differential pairs are set so that: (1) the output current saturates to a constant value when the applied voltage is out of the voltage range used to encode the external gray-scale images [1] and (2) the small signal transconductance gain of the differential pair, G_m, is greater than $1/R_x$, where R_x is the state resistor shown in Figure 3, as this is necessary to obtain weight gains greater than unity as required for most CNN templates [2]. In our case the voltage range used to encode gray-scale images is (−250mV, +250mV), and the G_m/R_x ratio is 3.5, which will allow a gain range of (−3.5, +3.5) for all the template weights, which is sufficient for most CNN operations reported in the literature.

Each one of the four-transistor stages on the right of Figure 5 can multiply their incoming differential current by multiplication factor between −1 and +1. The transistors are sized so that all multiplication factors in the entire range (−1, +1) can be obtained by appropriate values of the external control voltages $v_{a(k,l)}$ and $v_{b(k,l)}$. The final output current from each pair of multipliers is converted to single-ended and fed into the state resistors of one of the neighboor cells. Figure 6 shows the transistor implementation the state resistor of a cell.

4. Chip Layout

In a 0.8 μm technology, and using three levels of metals, each cell occupies an area of 232 μm × 263.5 μm. The cells are designed so they can be tiled together, as they include all the wiring for intercell connections, data loading, data output, control and bias. Figure 7 shows a micrograph of one cell with portions of the surrounding eight neighboor cells. Edge cells and the input/output circuit blocks are tiled at the borders of the processing array. Figure 8 shows a micrograph of the entire 5.5 mm × 4.7 mm silicon die.

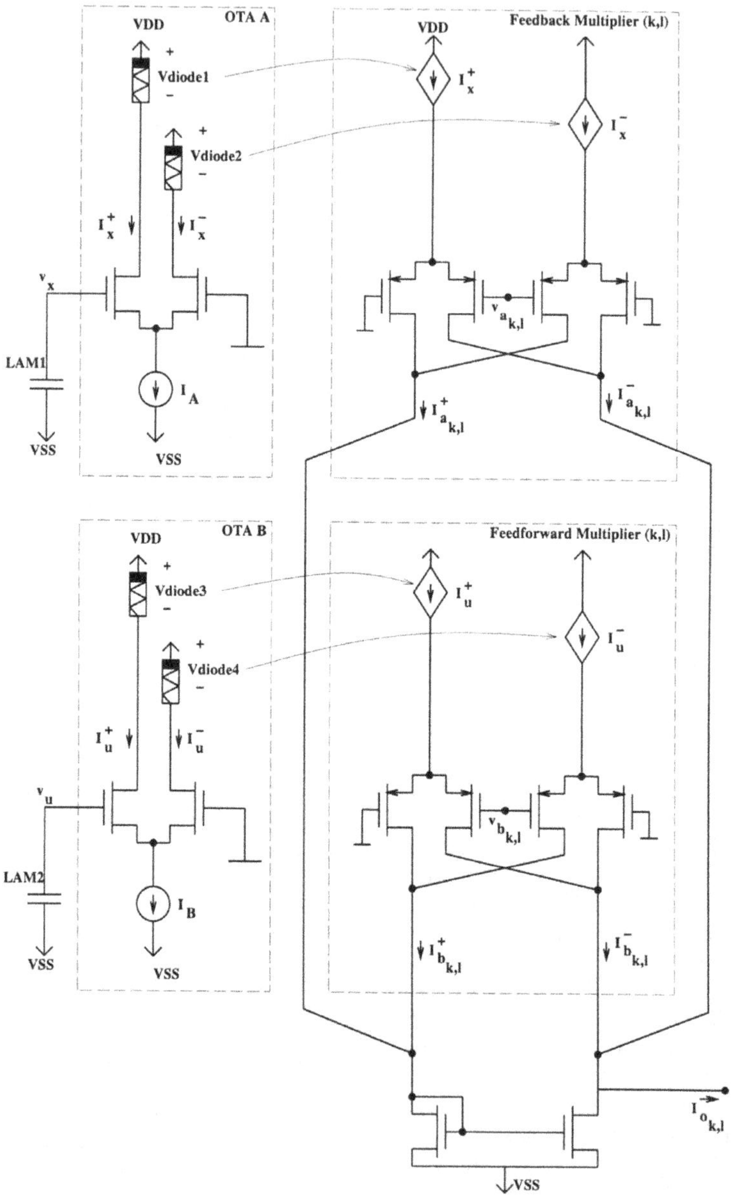

Fig. 5. Amplifiers and a pair of multipliers.

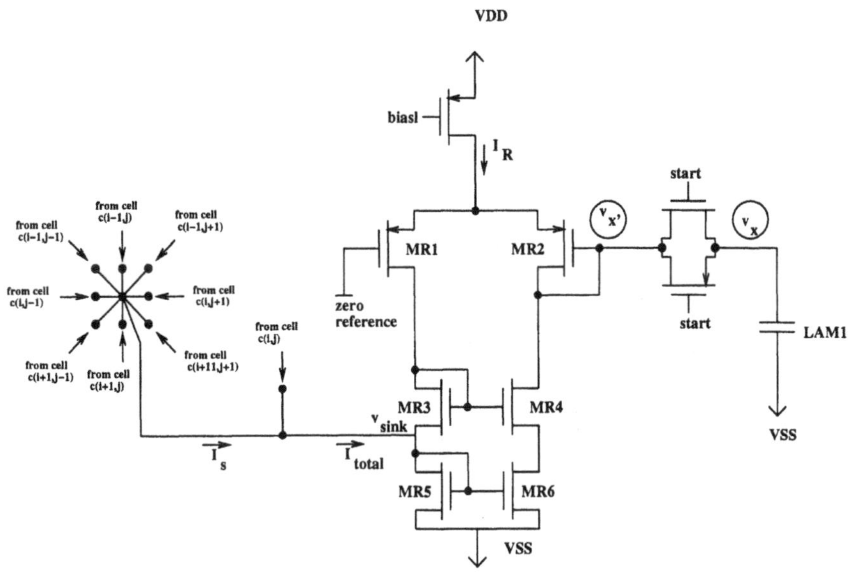

Fig. 6. State resistor.

5. Chip Operation

The chip reported in this paper has been interfaced to the CNN chip prototyping system built at the Hungarian Academy of Sciences [14]. This system down load images into the chip and to control its operation via a personal computer running a language called Analogic Machine Code [14]. The interface board with the chip was designed and built by Dr. T. Roska and Dr. A. Zarandy at the Hungarian Academy of Sciences according to the pinout and chip control sequences described in [4]. The tests were conducted by Dr. T. Roska, Dr. A. Zarandy, and the authors of this paper by using the complete CNN chip prototyping system interfaced with the chip.

The tests show correct operation for all basic functions of the chip, which include: (1) the loading of external image data and their storage in internal analog or logic memories, (2) the CNN analog and logic processing of stored images, (3) the local storage of intermediate processed images in the local memories, (4) the local retrieval of intermediate images for subsequent processing, and (5) the readout of final processed images using either the parallel or the serial output port. Examples of successful CNN applications [13] that have been tested are the following.

- connected component detection in 8 directions
- edge detection
- corner detection
- shadow detection
- halftoning
- hole filler
- diagonal line eraser
- the 16 boolean operations of two binary images
- classification of objects by their size

The speed of the cell dynamics have been measured by using the sensing mode of the chip, which allows to access, without disturbing the dynamics, the internal transient waveforms of any selected cell during the chip operation.

Figures 9 and 10 show an example of two connected component detection operations performed by the chip. This is a type of image feature extraction operation [5,13] used in the early stages of some handwritten character recognition algorithms. Figure 9 shows the case of vertical connected component detection, which detects the number of connected groups of black pixels in the vertical direction of the image. The image shown on the left is the input downloaded into the chip, the image on the right is the image readout obtained experimentally from the chip after its operation. Figure 10 shows a similar example with the chip reprogrammed for diagonal connected component detection, which detects the number of connected groups of black pixels

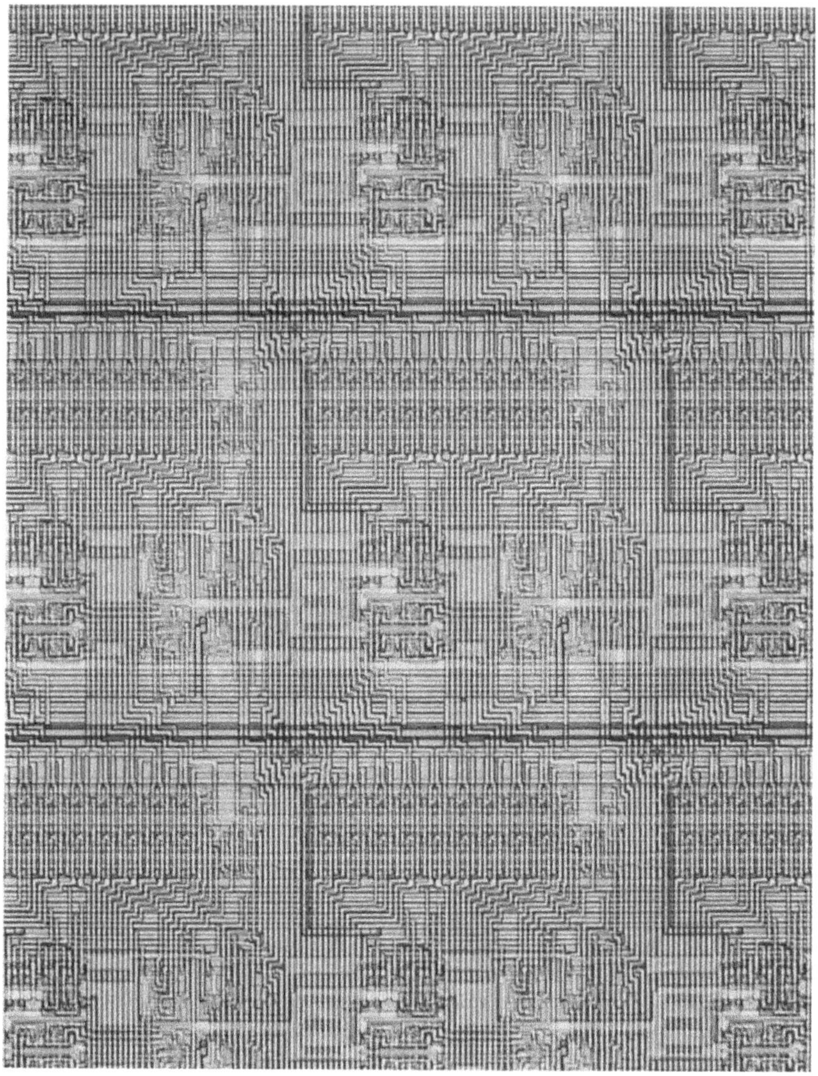

Fig. 7. Micrograph of one cell and part of the surrounding ones.

in one of the diagonal directions. The image shown on the left is the initial image which was downloaded into the chip and the image shown on the right is the new image readout from the chip. Both examples involve the propagation of information across the cell array and they show correct operation.

6. Conclusions

The chip presented in this paper is the first continuous-time CNN Universal chip combining the capability to input and output gray-scale images and with the capability to locally store and recall images.

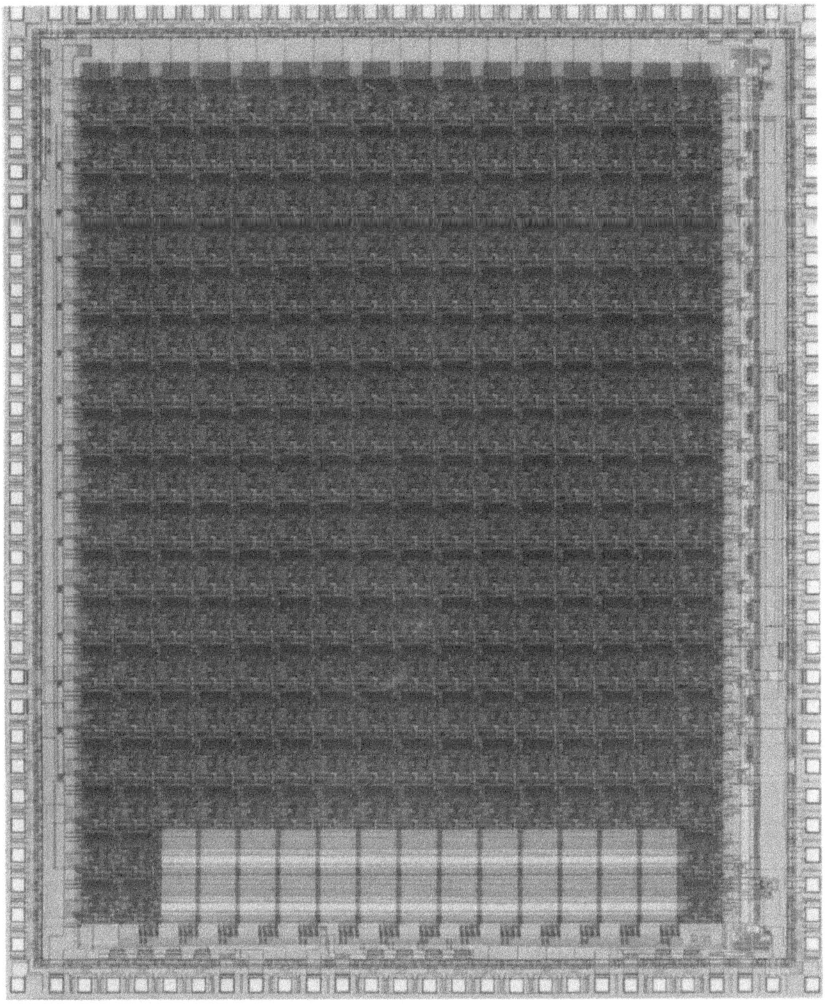

Fig. 8. Photomicrograph of the entire die.

The chip follows the original CNN dynamics presented in [1]. The chip contains 4,608 multipliers operating in parallel, the largest reported so far for a CNN chip, and implement the CNN equations with a $R_x \times C_x$ product under 90 ns, which is to our knowledge the fastest dynamics reported so far for a programmable CNN chip, providing a peak computing power (excluding I/O) of one billion pixels per second. An emulation of the dynamics of this chip would require about 200 billion arithmetic operations per second.

Acknowledgements

The authors acknowledge and thank Professor T. Roska and Dr. A. Zarandy for the developing of the prototyping system interface with this chip and for conducting many experimental tests of the chip.

This work was conducted at the University of California, Berkeley, sponsored by the Joint Services Electronics program under contract F49620-94-C-0038. The United States Government is authorized to repro-

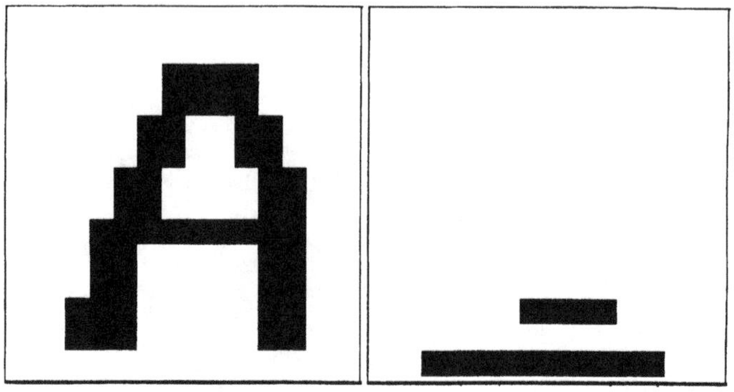

Fig. 9. Operation example: vertical connected component detector.

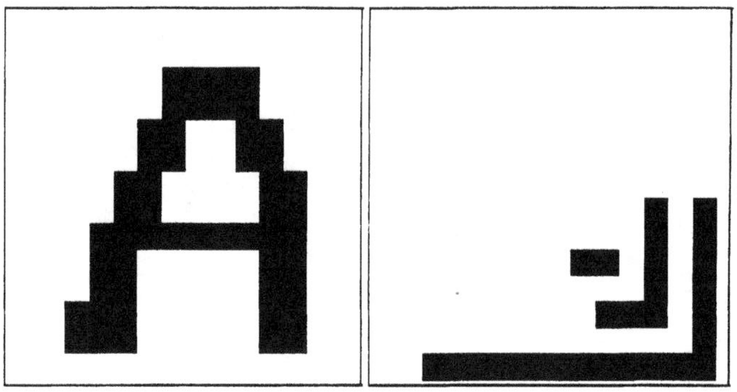

Fig. 10. Operation example: diagonal connected component detector.

duce and distribute reprints for governmental purposes notwithstanding any copyright notation hereon.

References

1. L. O. Chua and L. Yang, "Cellular neural networks: Theory." *IEEE Transactions on Circuits and Systems* 35, pp. 1257–1272, 1988.
2. L. O. Chua and L. Yang, "Cellular neural networks: Applications." *IEEE Transactions on Circuits and Systems* 35, pp. 1273–1290, 1988.
3. J. M. Cruz and L. O. Chua, "Design of high-speed high-density CNNs in CMOS technology," in T. Roska and J. Vandewalle (editors), *Cellular Neural Networks*, John Wiley and Sons: England, 1993.
4. J. M. Cruz, "VLSI implementation of cellular neural betworks," Ph.D. Thesis, University of California, Berkeley, 1996.
5. J. M. Cruz and L. O. Chua, "A CNN chip for connected component detection." *IEEE Transactions on Circuits and Systems* 38, pp. 812–817, 1991.
6. T. Roska and L. O. Chua, "The CNN universal machine: An analogic array computer." *IEEE Transactions on Circuits and Systems* 40, pp. 163–173, March 1993.
7. K. Crounse and L. O. Chua, "The CNN universal machine is as universal as a turing machine." *IEEE Transactions on Circuits and Systems* 43, pp. 353–355, April 1996.
8. R. Dominguez, S. Espejo, A. Rodriguez, and R. Carmona, "A CNN universal chip in CMOS technology," in *Third IEEE International Workshop on Cellular Neural Networks and their Applications*, December 1994, pp. 91–96.
9. H. Harrer, J. A. Nossek, T. Roska and L. O. Chua, "A current mode DTCNN universal chip," in *Third IEEE International Workshop on Cellular Neural Networks and their Applications*, December 1994, pp. 91–96.
10. M. D. Doan, M. Glesner, R. Chakrabaty, M. Heidenreich et al., "Realisation of a digital cellular neural network for image processing," in *Third IEEE International Workshop on Cellular Neural Networks and their Applications*, December 1994, pp. 85–90.
11. P. Kinget and M. S. Steyaert, "A programmable analog cellular neural network CMOS chip for high speed image processing." *IEEE Journal of Solid State Circuits* 30, pp. 235–243, March 1995.

12. J. M. Cruz and L. O. Chua, "A core cell and I/O circuitry of an analog-input dual-output CNN universal chip," in *Proceedings of the International Symposium on Nonlinear Theory and its Applications*, December 1995, pp. 661–666.
13. T. Roska and L. Kek (editors), "Analogic CNN program library," Analogic and Neural Computing Laboratory, Hungarian Academy of Sciences, Kende u. 13-17, H-1111, Budapest, Memorandum No. DNS-5-1994, June 1994.
14. T. Roska, P. Szolgay, A. Zarandy, P. Venetianer, et al., "On a CNN chip prototyping system," in *Third IEEE International Workshop on Cellular Neural Networks and their Applications*, December 1994, pp. 375–380.

Jose M. Cruz received his M.S. and Ph.D. degrees both in Electrical Engineering and Computer Sciences from the University of California at Berkeley in 1992 and 1996, respectively. Previously he received his B.S. degree in Physics from the University of Sevilla, Spain, and a Master of Engineering, with great distinction, from the University of Leuven, Belgium.

He has been a researcher at the ESAT Research Laboratories of the University of Leuven at Heverlee, Belgium, and at the Electronics Research Laboratory of the University of California at Berkeley. In 1991, while at Berkeley, he developed the first successful implementation of a CNN chip.

Dr. Cruz is currently with Sun Microsystems, Inc., California.

Leon Chua is currently a Professor of electrical engineering and computer sciences at the University of California, Berkeley. His research interests are in the areas of general *nonlinear network and system theory*. He has been a consultant to various electronic industries in the areas of nonlinear network analysis, modeling, and computer aided design. He is the author of *Introduction to Nonlinear Network Theory* (McGraw-Hill, 1969), and a coauthor of the books *Computer Aided Analysis of Electronic Circuits: Algorithms and Computational Techniques* (Prentice-Hall, 1975), *Linear and Nonlinear CIrcuits* (McGraw-Hill, 1987), and *Practical Numerical Algorithms for Chaotic Systems* (Springer-Verlag, 1989). He has published many papers in the area of nonlinear networks and systems. He served as editor of the IEEE *Transactions on Circuits and Systems* from 1973 to 1975 and as President of the *IEEE Society on Circuits and Systems* in 1976. He is presently the editor of the International Journal of Bifurcation and Chaos and a deputy editor of the *International Journal of Circuit Theory and Applications*.

Professor Chua is the holder of six U.S. patents. He is also the recipient of several awards and prizes, including the 1967 IEEE Browder J. Thompson Memorial Prize Award, the 1973 IEEE W.R.G. Baker Prize Award, the 1974 Frederick Emmons Terman Award, the 1976 Miller Research Professorship, the 1982 Senior Visiting Fellowship at Cambridge University, the 1982/83 Alexander von Humbolt Senior U.S. Scientist Award at the Technical University of Munich, the 1983/84 Visiting U.S. Scientist Award at Waseda University, Tokyo, the IEEE Centennial Medal in 1985, the 1985 Myril B. Reed Best Paper Prize, the 1985 and 1989 IEEE Guillemin-Cauer Prizes, and the 1995 M.E. Van Valkenburg Award. In 1986 he was awarded a *Professeur Invité* International Award at the University of Paris-Sud from the French Ministry of Education. He was also awarded a *Doctor Honoris Causa* from the Ecole Polytechnique Fédérale-Lausanne, Switzerland, in 1983, an Honorary Doctorate from the University of 1992, a *Doctor Honoris Causa* from the Technical University of Budapest, Hungary, in 1994, a *Doctor Honoris Causa* from the University of Santiago de Compostela, Spain, in 1995, and a *Doctor Honoris Causa* from the University of Frankfurt, Germany, in 1996.

Analog Integrated Circuits and Signal Processing, 15, 239–250 (1998)

A 6 × 6 Cells Interconnection-Oriented Programmable Chip for CNN

MARIO SALERNO, FAUSTO SARGENI AND VINCENZO BONAIUTO

University of Rome - Tor Vergata, Department of Electronic Engineering, Via della Ricerca Scientifica, 1 - 00133 ROME - ITALY

sargeni@utovrm.it

Received August 8, 1996; Accepted March 4, 1997

Abstract. The implementation of a versatile VLSI chip certainly represents an important step to improve the research on Cellular Neural Networks. In this paper a VLSI realization of the multi-chip oriented, the 6 × 6 Digitally Programmable Cellular Neural Network (6 × 6DPCNN) chip, will be presented. This chip covers most of the available one-neighbourhood templates for image processing applications. Moreover, it can be easily interconnected to others to carry out very large CNN arrays. The designs and some measured results of a single chip and a multi-chip board (the 720 DPCNN System) will be shown.

1. Introduction

Cellular Neural Networks (CNN's), introduced by L. O. Chua and L. Yang in 1988 [1], represent a remarkable improvement in hardware implementation of Artificial Neural Networks. In fact, their regular structure and particularly the local connectivity feature makes this class of neural networks really appealing for VLSI implementations. The CNN are widely used in several application fields, such as image processing and pattern recognition [2–3] because of the continuous-time dynamic system feature which allows a parallel real-time signal processing.

In recent years, several hardware implementations were proposed in several papers [4–10] with different circuital solutions and partial/full programmability.

It is worth to note that the hardware realization of a fully programmable CNN machine represents an essential step in developing of this research field. In particular, the main key feature in the implementation of the templates programmability is represented by the realization of selectable Voltage Controlled Current Sources (VCCS).

In this class of programmable CNN implementations, the authors presented a fully digitally programmable CNN chip with 3 × 3 cells (3 × 3DPCNN chip). This chip, manufactured and successfully tested [14–16], is made up of nine neural cells placed in a 3 × 3 grid. Moreover, its programmability feature has been carried out by using the Digitally Programmable Transconductance Amplifier (DPTA) [11]. In fact, by this device you can carry out a VCCS whose control parameter is digitally selectable.

In this paper the new 6 × 6DPCNN chip will be presented. Similarly to the 3 × 3DPCNN chip, the 6 × 6DPCNN chip has been designed as well by using a current-mode technique to obtain an interconnection-oriented circuit. Moreover, it is fully controllable by a digital processor that allows you to select the Cloning Templates as well as the input patterns. The chip has been successfully tested and the measurements that will presented in this paper concern a single chip or the 720 DPCNN System, a multi-board system with twenty chips connected together.

2. Recall of the CNN State Equations

In order to describe the chip design, let us recall the state equations of the CNN [1]

$$C\frac{dv_x^{(i,j)}(t)}{dt} = -\frac{1}{R}v_x^{(i,j)}(t) + f^{(i,j)} + g^{(i,j)} \qquad (1)$$

where the terms $f^{(i,j)}$ and $g^{(i,j)}$, defined as follows,

$$f^{(i,j)} = \sum_{C(k,l)\in N_r(i,j)} A(i,j;k,l)v_y^{(k,l)}(t) \quad (2)$$

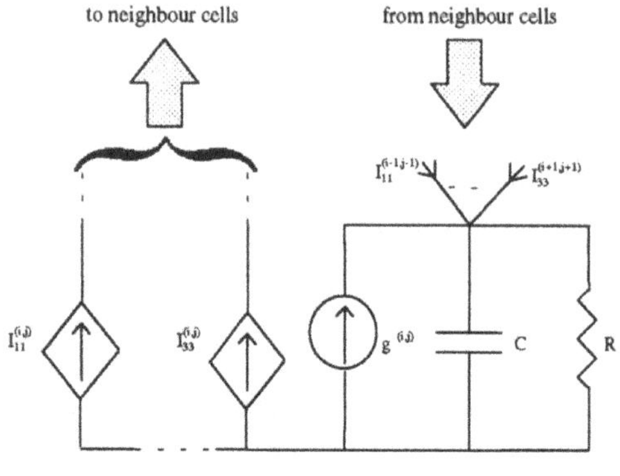

Fig. 1. Modified scheme of the CNN cell.

and

$$g^{(i,j)} = \sum_{C(k,l)\in N_r(i,j)} B(i,j;k,l)v_u^{(k,l)} + I \quad (3)$$

are current contributions and therefore the templates **A** and **B** are transconductance parameters and the output voltages $v_y^{(i,j)}$ are defined as

$$v_y^{(k,e)} = \frac{1}{2}[|v_x^{(k,l)} + 1| - |v_x^{(k,l)} - 1|] \quad (4)$$

The above state equation have to be properly modified to carry out a suitable VLSI implementation using a current-mode approach as proposed in previous designs and papers [4], [5].

On this purpose, let us focus our attention on the term $f^{(i,j)}$. If we include the effect of the non linearity (4) in the equation (2) we obtain the following expression

$$I_{mn}^{(i,j)} = \frac{a_{mn}}{2}[|v_x^{(i,j)} + 1| - |v_x^{(i,j)} - 1|]$$
$$\text{with } m,n = 1,2,3 \quad (5)$$

Following the current-mode approach, these current contributions can be generated inside each cell and carried directly to its neighbourhood. In such way, each cell receives from the neighbours only current contributions instead of voltages and, at the same time, gives them its own current contributions. For example, the cell (i, j) will receive from the cell $(i-1, j-1)$ the contribution $I_{11}^{(i-1,j-1)}$ and give the contribution $I_{33}^{(i,j)}$ to the same cell. In Fig. 1 is depicted this modified scheme.

The implementation of the current contributions of eq. (5) can be carried out by the Digitally Programmable Transconductance Amplifier (DPTA) [11]. As shown in Fig. 2, the DPTA architecture consists of an array of four OTA's, operating in the same linear region whose transconductance parameters are weighted as power of two (Table 1). Current mirrors are used to obtain the negative and positive differential output currents. By means of a digital control word, in this design of five bits, you are able to select the proper OTAs configuration to obtain the required current value.

By this device, a dynamic range of the transconductance parameter of $[-75, +75]$ μA/V with a step of 5.0μA/V in a linear region of 200mV (absolute value), can be achieved. Consequently the output current ranges in $[+7.5, -7.5]$ μA with a step of 0.5 μA which means a range of the templates of $[-7.5, +7.5]$.

Let consider the term $g^{(i,j)}$. We can note that this term is time-independent and then we chose to compute these calculations $g^{(i,j)}$ by a host microprocessor. These results, one for each cell, can be used to program a linear current source inside each cell. To implement this programmable current source, we already can use a DPTA. In fact, it can also be seen as a digitally controlled current source by fixing the input voltage and selecting the output current by the control word. To cover most of the application fields, the resolution of the DPTA used to implement the term $g^{(i,j)}$ has been improved by one bit (from 5 to 6 bits). This bit, i.e. a further OTA and related switches, has been added to the least significative section of the DPTA. In this way, a fine resolution of 0.25μA in the range $[-7.75, +7.75]$ μA has been achieved for the output current. This means that the term $g^{(i,j)}$ can range from -7.75 to $+7.75$. By this improvement we can use this VLSI implementation of CNN in a very large application field. Even under the worst input pattern conditions, this chip can deal with more than 80% of one-neighbourhood fixed template applications reported in [3].

3. VLSI Architecture of a 6 × 6DPCNN Chip

The 6 × 6 DPCNN chip is a fully programmable one-neighbourhood CNN with modular capability. In fact, by this chip you can implement wide CNN arrays simply connecting more of them together. Moreover it was specially designed with a digital bus, so to be easily controlled by a Personal Computer. In this way, a CNN system, made up by the CNN chip itself and the PC, represents a really powerful tool which enables you to select the templates to experimentally discover new CNN algorithms. Furthermore, the chip will allow a

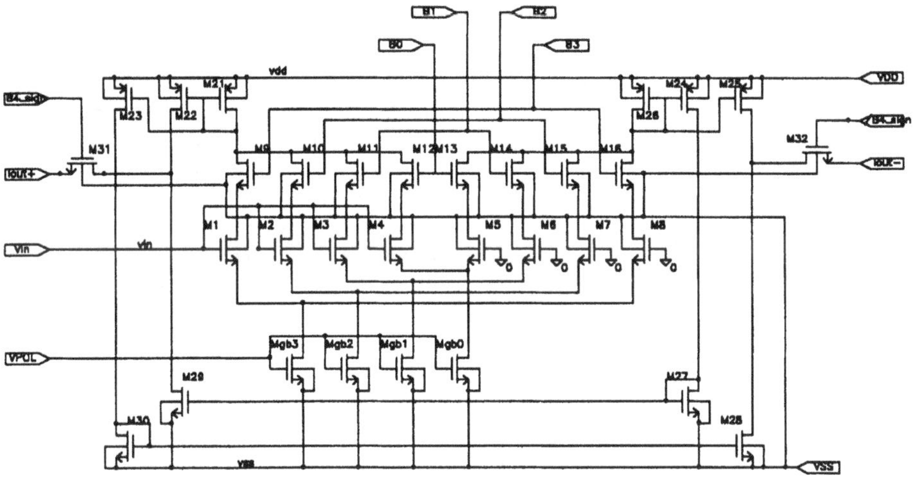

Fig. 2. Electrical scheme of the Digitally Programmable Transconductance Amplifier.

Table 1. Parameters of the OTAs.

OTA#	g_m	Differential Pairs Mos, W, L	Current Generators Mos, W, L, W/L
0	5 μA/V	M4-M5, 2.8, 3.0	Mgb0, 2.8, 25.0, 0.112
1	10 μA/V	M3-M6, 4.6, 2.4	Mgb1, 2.8, 13.2, 0.212
2	20 μA/V	M2-M7, 9.2, 2.6	Mgb2, 5.4, 12.8, 0.421
3	40 μA/V	M1-M8, 19.8, 2.6	Mgb3, 10.8, 12.8, 0.843

wide range investigation on the dynamic behaviour of a real CNN network. On this purpose, it is worth to note that the special architecture used to implement the neural cell does not modify the structure of the Chua's original model at all.

In recent papers [12–17], the authors have presented the 3 × 3DPCNN chip. This chip was the first example of VLSI implementation of an interconnection oriented CNN chip. Following the same criterion, the 6 × 6DPCNN chip has been designed, manufactured and tested. In Fig. 3 is depicted the block scheme of this chip.

In this figure you can distinguish the two main sections of the chip: the analogue CNN processing area and the digital local memory area.

The analogue area is made up of 36 neural cells arranged in a two dimensional grid. Owing to the successful experience of the former 3 × 3 realization, we made use of the same layout of the analogue core as

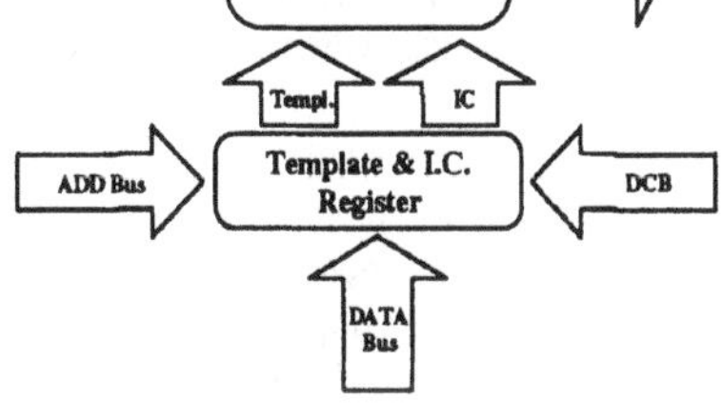

Fig. 3. Block scheme of the 6 × 6DPCNN chip.

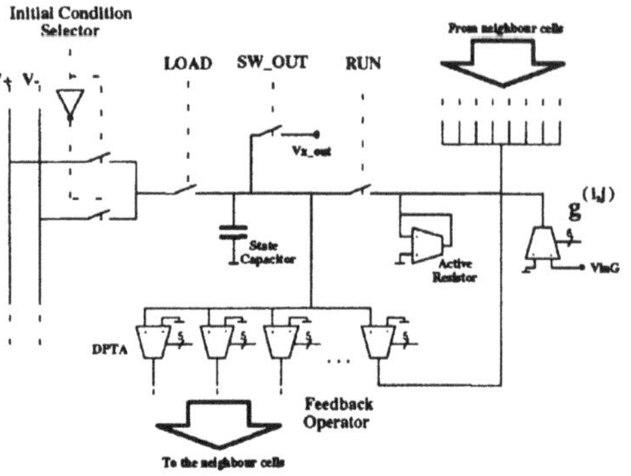

Fig. 4. Scheme of a single cell in the 6 × 6DPCNN chip.

the basic building block for the new project. So, four of these 3 × 3 cells building blocks have been properly interconnected to obtain the new 6 × 6 cells structure. Similarly to that implementation, each cell is composed of nine 5-bit DPTA's for the entries of template **A** and one 6-bit DPTA for the term $g^{(i,j)}$. Therefore, each cell (Fig. 4) is made up of ten DPTA's, a capacitor of nominal value of 1pF (metal1-poly-bulk) and an active resistor (negative-feedback OTA) having a nominal value of 170 kΩ (layout extracted values). In this way, it is possible to control each cell of the 6 × 6DPCNN by 45 bits for the template **A** (i.e. 5 bits for each of the 9 entries) and 6 bits for the term $g^{(i,j)}$. As a result, the chip is controlled by 261 bits register: one 45 bits register (A Register) for the template A plus a 216 bits register (G Register) for the terms $\mathbf{g}^{(i,j)}$ (i.e. 6 bits for each of the 36 cellular cells). Moreover, a further register (IC Register) made up of 36 bits (i.e. one bit for each cell) has been added to this digital local memory to allow the selection of the initial state conditions. In fact, in our design we choose to select saturated values only for the input patterns, so, we need two different voltage values only for the black and white input, respectively. Each output of the IC Register will drive a 2:1 analogue multiplexer, which enables the state capacitor to be charged to a positive or negative initial condition. By using such a technique, it will be possible, in a future chip, to implement a wider set of initial condition values simply increasing the dimension of the IC Register and the analogue multiplexers.

Therefore, this 297 bits local memory allows the template and initial conditions storing operation controlled by a Personal Computer. In fact, this memory, addressed by six lines, is connected to an 8-bit external data bus. Moreover, two different buses permit the chip control: the Digital Control Bus (**DCB**) and the Analogue Control Bus (**ACB**). In particular, the **DCB** is composed of the chip select signal (**/CS**), a strobe line (**/IOW**) for the loading operations and three reset lines (**RESET_A, RESET_G, RESET_IC**) to reset the template A and current contributions $g^{(i,j)}$ and the initial conditions entries respectively. The **ACB** is composed of the **LOAD**, the **RUN** and the **SW_OUT** signals. The first signal enables the initial voltages to be stored in the state capacitors, the **LOAD** signal starts up the analogue processing of the whole network, while the latter signal enables the read operation of the analogue state voltages.

The outputs of the network are represented by the 36 state voltage values of the 36 cells of the chip. On this purpose, these voltages have been led to 36 external pads to allow a more reliable chip testing as well as an investigation on the dynamic behaviour of the network.

A common problem in CNN chips, as well as in parallel data processing, certainly is the speed bottleneck represented by the I/O operations time. So, it is strongly important to significantly reduce the amount of those operations. By using the 6 × 6 DPCNN chip, you are able to perform multistep algorithms by using as new initial conditions the steady-state voltages of the previous evolution. In fact, by turning off the LOAD and RUN switches, you electrically insulate the state capacitor from the circuit. So, the state capacitor will maintain its own charge for the time required to the loading task of the successive Template. In this way, you can perform a complex CNN algorithm without output and re-input the processing partial results.

Similarly to the 3 × 3DPCNN chip design, this implementation was designed to achieve a modular architecture as well. As a matter of fact, the 28 current contributions of the twenty boundary cells have been led to external pads in order to carry out a modular and expandable architecture. Although this is true, in the 3 × 3DPCNN chip each current contribution was led to a different external pad to make more reliably the testing task. Conversely, in this new implementation, the current contributions directed to the same input of the adjacent chip have been connected together inside the chip in order to bound the total number of pads.

In addition, the bound in the number of the pads allows the die area to be limited. Undoubtedly, this interconnection-oriented approach, in this direct form, presents the intrinsic limit that the die area is constrained by the number of the pads rather than the inter-

Fig. 5. Microphotography of the 6 × 6DPCNN chip.

nal core size. To illustrate this, if we consider a network with $N \times N$ cells, the number of the interconnections for the current contributions is given by $4 \cdot (N + 1)$. In addition, further $4 \cdot (N - 1)$ pads are necessary for the current inputs of the periphery cells with a total of $8 \cdot N$ pads. For example, if we consider a future 32 × 32DPCNN chip, it will require 256 pads for the interconnections only!

This limit can be surely overcome in a new version of the chip by a better interconnection strategy that is in advanced development status.

The 6 × 6DPCNN chip die area is 7108 × 7314μm^2 by 1.5μm CMOS process, the analogue core represents the 23% only of the total area. The chip is packaged in a 144 PGA. The microphotography is shown in Fig. 5.

Some static measurements have been carried out in order to characterize the chip in terms of power consumption and accuracy of the template's implementation. As regard first point, the chip shows a power consumption of about 375mW for the digital section, reaching, for the analogue cells, a maximum value of 13.36mW with a standard power supply of 5.0V. Later, in order to evaluate the accuracy of the template's implementation, many tests have been carried out on different DPTA located in various positions inside the chip. All these measures show a minimum spread with respect of the medium values.

4. Test Results

The functionality of the chip has been tested using several CNN templates in every input configuration with successful results. The tests were performed in the single-chip board and the 720 DPCNN System, where 20 chips were connected together.

These boards were fitted with a proper circuitry to interface the AT-bus of a Personal Computer. In this way, you can control the CNN by means of a dedicated software tool running on the PC. In particular, this software allows you to control the board enabling the selection of the entries of the templates ***A***, ***B***, the bias ***I*** and the initial conditions values. Moreover, you are able to digitally acquire the saturated steady-state output voltages of the network or reuse them to perform a multistep CNN algorithms as well.

In order to evaluate the processing performances of the CNN hardware, let define two operative quantities

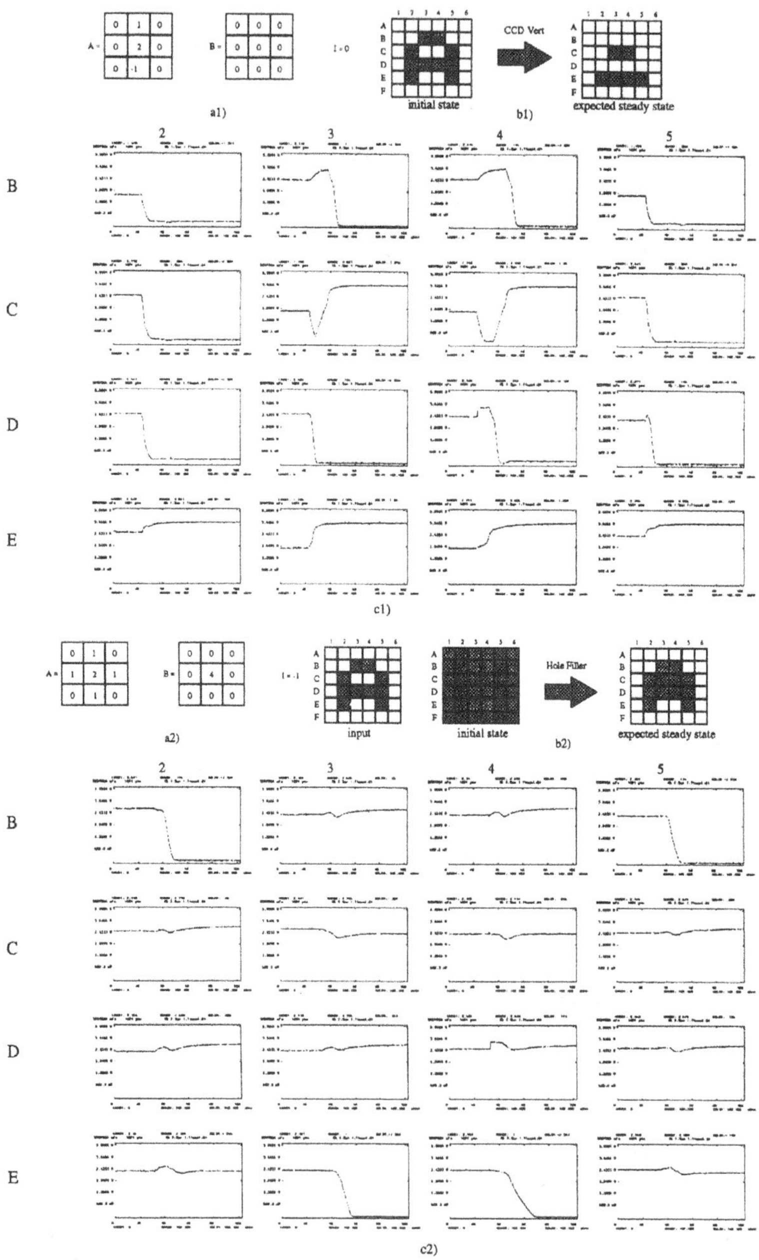

A =
B =
I = 0
CCD Vert
initial state
expected steady state
a1)
b1)
c1)
A =
B =
I = -1
Hole Filler
input
initial state
expected steady state
a2)
b2)
c2)
2
3
4
5
B
C
D
E

Fig. 6.

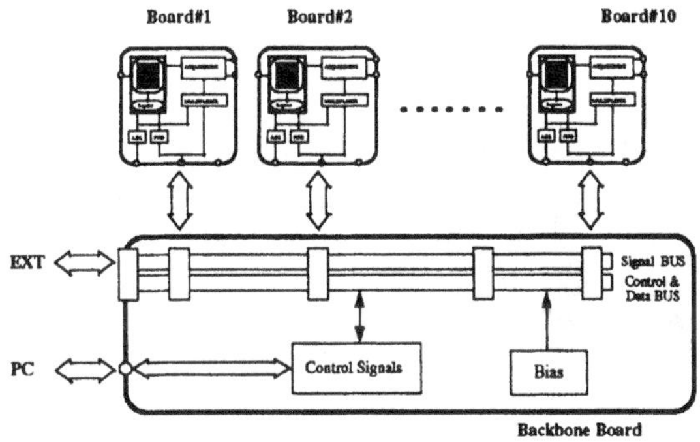

Fig. 7. Scheme of half 720 CNN System.

Fig. 8. Photography of the 720 CNN System.

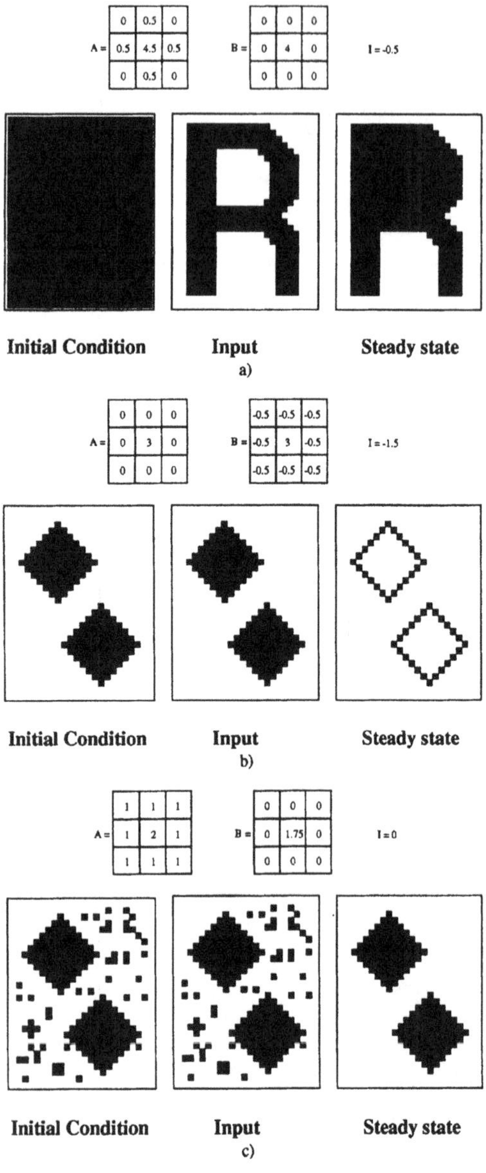

Fig. 9.

for the measurement task, the *operation time* and the *convergence time*.

Definition 1 In a hardware CNN cell $C(i, j)$, the *operation time*, is the interval of time (if exist) between the start of the transient and the instant such that

$$|v_x^{(i,j)}(t_o^{(i,j)})| = 1 \text{ and } |v_x^{(i,j)}(t)| > 1 \text{ for } t > t_o^{(i,j)}.$$

Definition 2 In a hardware CNN cell $C(i, j)$, the *convergence time* is the interval of time (if exists) between the start of the transient and the instant such that

$$|v_x^{(i,j)}(t_c^{(i,j)}) - a^{(i,j)}| = 0.1 \cdot a^{(i,j)}$$
$$\text{and } |v_x^{(i,j)}(t) - a^{(i,j)}| < 0.1 \cdot a^{(i,j)} \quad \text{for } t > t_c^{(i,j)}$$
$$\text{with } a^{(i,j)} = \lim_{t\to\infty} |v_x^{(i,j)}(t)|.$$

It is worth to note that, for the whole CNN hardware, the important quantities are $T_0 = \max\{t_o^{(i,j)}\}$ and the range of $t_c^{(i,j)}$. These measurements will be given in the examples shown below.

Among the performed tests, the "***HOLE FILLER***" and "***CCD VERTICAL***" operations will be shown below as an example of the single-chip board and "***HOLE FILLER***," "***EDGE DETECTION***" and "***SMALL KILLER***" operations will be shown in detail as an example of the functionalities of the 720 DPCNN System.

Regarding the "***HOLE FILLER***" operation, the input must be stored as Input voltage V_u while the state voltage V_x should be black in the whole chip. Conversely, in "***CCD VERTICAL***" operation, the input values must be stored as initial state voltages V_x and there are no instructions about the V_u values (actually the **B** matrix is identically equal to zero).

Fig. 6 shows the measured transient waveforms relative to that operations. In particular, the section **a** shows the selected templates, the section **b** the inputs and the expected steady states, and, finally, the section **c** the transient state voltage waveforms of the 16 inner cells. In this figure, you can observe the correct behaviour of the chip. Moreover, you can note that, for the specific templates reported in sections **a**, the convergence time, $\mathbf{t_c}$, ranges from 14.2μs to 39.5μs, and the operation time, $\mathbf{T_o}$, is 13.9μs for the "***HOLE FILLER***." Regarding the "***CCD VERTICAL***" operations, $\mathbf{t_c}$ ranges from 4.5μs to 29.6μs, and $\mathbf{T_o}$ is 27.4μs.

The 720 DPCNN System has been implemented by connecting together twenty "*single-chip board*" where each 6 × 6DPCNN chip was assembled with a proper circuitry for controlling, buffering and carrying out the binary conversion of the steady states of the network.

In particular, ten of these boards have been connected in a grid of 5 × 2 chips (30 × 12 CNN cells) onto a

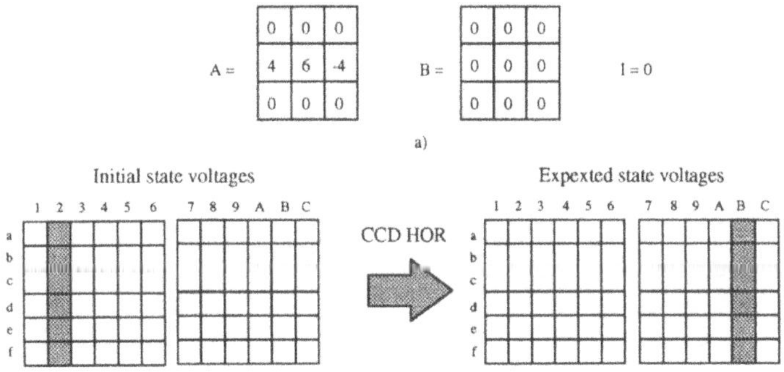

Fig. 10. (1) CCD Hor operation (a) Template (b) Inputs and expected steady state.

"*backbone board*" as shown in Fig. 7. Moreover these backbone boards can be connected together to obtain a more complex equipment. On this purpose, each of the backbone boards has been designed in order to allow the connection with others. In this way, simply connecting the backbone boards, you are able to implement a classic plane CNNs (sized 30 × 24 cells or 60 × 12 cells) or non-conventional CNN topologies. This neural equipment represent a useful tool in studying of new CNN operation as well as the dynamic behaviour of the network. The Backbone Boards as well as the Single Chip Board are double layer PCBs with size of 431 × 263 mm^2 and 160 × 230 mm^2 respectively. The total power comsumption is 40W at 5V. The photography of the 720 DPCNN System is shown in Fig. 8.

The functionality of the 720 DCNN System has been tested with several operation from [3]. All the measurements gave correct results. In particular, Fig. 9a depicts the Hole Filler Operation: this operation, in fact, fills all the holes existing in the input pattern. By using the Template shown in the Figure, the convergence time, $\mathbf{t_c}$, ranges from 1.4μs to 295.6μs, and the operation time $\mathbf{T_o}$ is 282.2μs.

In Fig. 9b, the *Edge Extraction* Operation is depicted. In fact, using the template shown in the Figure, the network detects the edge of the input pattern. In this example, the convergence time, $\mathbf{t_c}$, ranges from 103 to 145.2μs. The operation time $\mathbf{T_o}$ is 102μs.

Finally in section c of the Fig. 9, template and input output data are shown for the *Small Killer* operation with a convergence time of 88 to 116 μs and an operation time $\mathbf{T_o}$ is 93.8μs.

In order to evaluate the delay due to the non uniform capacitances, an example of a "***CCD HORIZONTAL***," with two chips connected together, a 6 × 12 CNN, will be shown below. Fig. 10 shows, in section **a**, the selected templates and, in section **b**, the inputs and the expected steady states. In Fig. 11 the relative transient waveforms are shown. In this example we choose as input pattern a single column of black pixels placed in second position. This column, because of the CCD template, has to propagate toward the right direction. As noted in Chua and Matsumoto [19], we can observe the particular behaviour as an " analogue shift register" of the network. In fact, each "positive pulse" is propagated toward right and settle down to a high value next a low value. The selected input pattern is capable to well show the different settling time of the single cells in a two chip case. This is a kind of "*race*" among the six rows of the CNN.

As shown in the Figures, the propagation of the "analogue pulse" is delayed when a parasitic capacitance is connected to the state node of a particular cell. Certainly, this always happens when the cell is placed on the boundary of the chip. Hence, the *propagation speed* of the "analogue pulse" in the first and last rows are slower than in the inner ones. Moreover you can observe a further delay effect when the analogue pulses cross the boundary of the chips. In this design, each cell of the border rows is connected to an external pad and, because of this, the value of state capacitor increases by an extra parasitic capacitance due to the pad and the printed circuit board.

By using a template like A = [4 6 − 4], B = [0], I = 0, the measured values show a difference of 89 μs between the *analogue pulses* propagating in the boundary

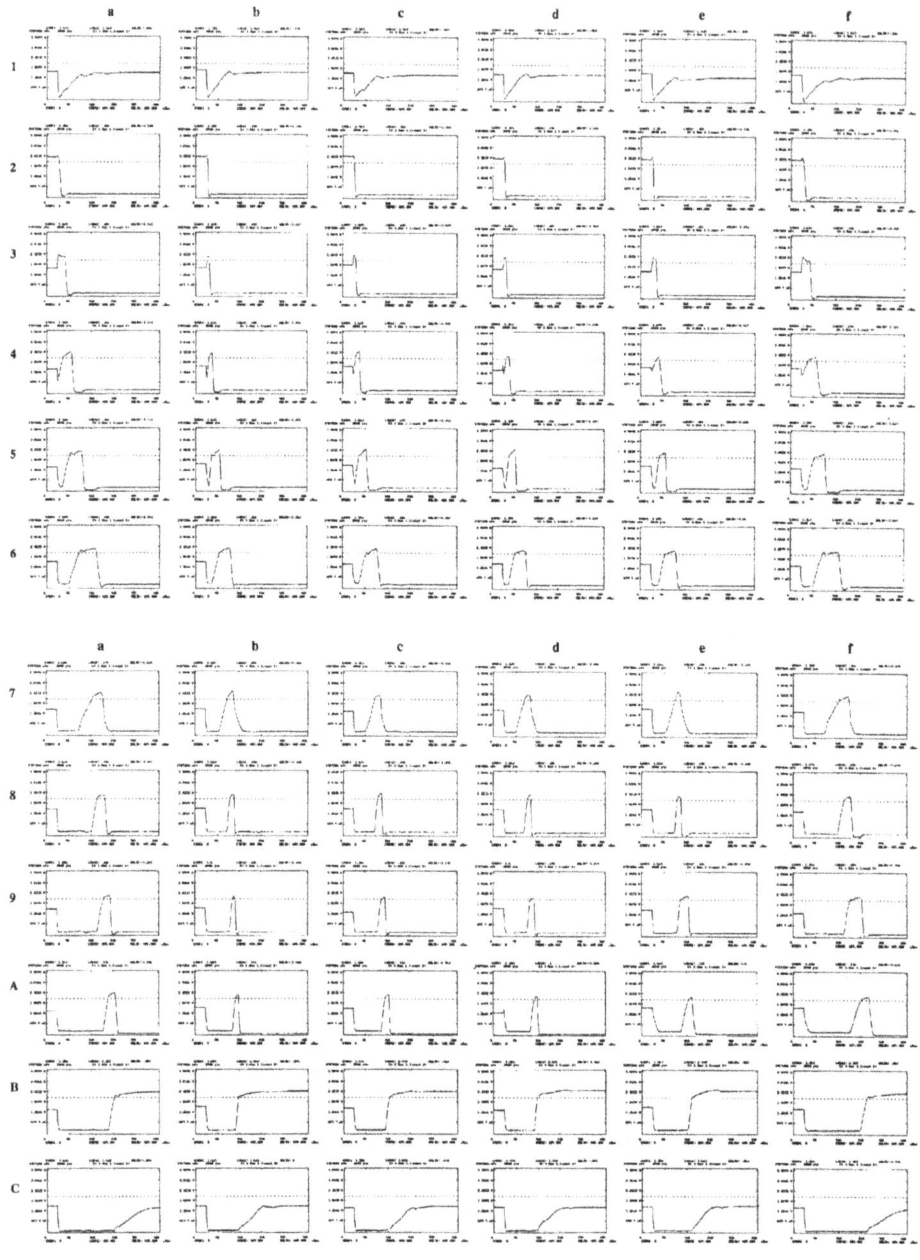

Fig. 11.

rows (i.e. row **a** and row **f** in Fig. 11) with respect of which propagating in the inner rows (i.e. rows from **b** to **e** in the same figure).

Although this delay in the propagation, the functionality has not been modified.

The correctness of the functionality of this CNN with non-uniform capacitance (i.e. non-uniform time constants) has been confirmed with all the other tested templates. These measures do not claim to demonstrate any theoretical issue, on the other hand, the tested templates represent different class of interactions between cells of the same neighbour. Anyway, any practical realization of ideal CNNs, both high integration single-chip or low integration multi-chip, surely implements a CNN with non-uniform time-constant and template.

5. Conclusions

In this paper the 6 × 6 Digitally Programmable CNN chip with discrete templates has been presented. This chip can deal with most of the available one-neighbourhood constant templates. Together with its fully digital programmability, a very powerful capability of this chip is its modularity that means that very large CNN array can be made up of several interconnected 6 × 6DPCNN chip. To test the interconnection feature of this chip, a twenty chips system has been designed, manufactured and tested.

References

1. L. O. Chua and L. Yang, "Cellular neural metworks: Theory and applications." *IEEE Trans. Circuits and Systems* 32, pp. 1257–1290, Oct. 1988.
2. T. Roska and J. A. Nossek (Eds), "Special issue on cellular neural networks." *IEEE Trans. on Circuits and Systems* 40(3), March 1993.
3. "Analogic CNN program library," Analogic and Neural Computing Laboratory, Computer and Automation Institute, Hungarian Academy of Sciences, Version 6.2, DNS-7-1995, April 1995.
4. J. M. Cruz and L. O. Chua, "A CNN chip for connected component detection." *IEEE Trans. Circuits and Systems*, 38, pp. 812–817, July 1991.
5. A. Rodriguez-Castro, S. Espejo, R. Dominguez-Castro, J. Huertas, and E. Sanchez-Sinencio, "Current-mode techniques for the implementation of continuous and iscrete-time cellular neural networks." *IEEE Trans. on Circuits and Systems-II* 40, pp. 147–155, March 1993.
6. K. Halonen, V. Porra, T. Roska, and L. O. Chua, "Programmable analogue VLSI CNN chip with local digital logic." *Interantional Journal of Circuit Theory and Applications* 20(5), pp. 573–582, 1992.
7. M. Anguita, F. J. Pelajo, A. Prieto, and J. Ortega, "Analog CMOS implementation of a discrete time CNN with programmable cloning templates." *IEEE Trans. on Circuits and Systems* 40(3), pp. 215–218, 1993.
8. P. Kinget and M. Steyaert, "A Programmable analog cellular neural network CMOS chip for high speed image processing." *IEEE Journal of Solid-State Circuits* 30(3), March 1995.
9. A. Paasio, A. Dawidziuk, and V. Porra, "High speed CNN VLSI implementation," in *Proceedings of the 1996 IEEE International Symposium on Circuits and Systems*, Atlanta, USA, May 1996, pp. 519–522.
10. E. Espejo, R. Carmona, R. Dominguez-Castro, and A. Rodriguez-Vazquez, "A CNN universal chip in CMOS technology." *Interantional Journal of Circuit Theory and Applications* 23, pp. 93–109, Jan.–Feb. 1996.
11. F. Sargeni, "Digitally programmable transconductance amplifier for CNN applications." *Electronics Letters* 30(11), pp. 870–872, May 1994.
12. F. Sargeni and V. Bonaiuto, "High performance digitally programmable CNN chip with discrete templates," in *Proceedings of CNNA-94, Third IEEE Int. Workshop on Cellular Neural Networks and their Applications*, Rome, Italy, Dec. 1994, pp. 67–72.
13. M. Salerno, F. Sargeni, and V. Bonaiuto, "DPCNN: a modular chip for large CNN arrays," in *IEEE International Conference on Circuits and Systems (ISCAS-95)*, Seattle, Washington, USA, May 1995, pp. 417–420.
14. F. Sargeni and V. Bonaiuto, "A fully digitally programmable CNN chip." *IEEE Trans. on Circuits and Systems-II* 42(11), pp. 741–745, Nov. 1995.
15. F. Sargeni and V. Bonaiuto, "A 3 × 3 digitally programmable CNN chip." *Interantional Journal of Circuit Theory and Applications* 24(3), pp. 369–379, 1996.
16. M. Salerno, F. Sargeni, and V. Bonaiuto, "A 9 × 9 multichip CNN board for cellular neural networks," in *Proceedings of CNNA-96, 4th IEEE International Workshop on Cellular Neural Networks and their Application*, Seville, Spain, June 1996, pp. 261–266.
17. M. Salerno, F. Sargeni, and V. Bonaiuto, "6 × 6DPCNN: A programmable mixed analogue-digital chip for cellular neural networks," in *Proceedings of CNNA-96, 4th IEEE International Workshop on Cellular Neural Networks and their Application*, Seville, Spain, June 1996, pp. 451–456.
18. T. Matsumoto, L. O. Chua, and H. Suzuki, "CNN cloning template: Connected component detector." *IEEE Trans. on Circuits and Systems* 37(5), pp. 633–635, May 1990.

Mario Salerno (born in Rome, 1944) is professor at the Engineering Faculty of the University of Rome at Tor

Vergata, Rome, Italy. Since 1970, his activity has dealt with circuit theory and applications, included various aspects in the fields of signal processing and digital technology. In the late Eighties, he was involved in various projects in the field of VLSI chip design and realization. At this purpose, he realized the Circuit Laboratory of the University, a laboratory with digital and analog design capabilities for ASIC development. At present, he is involved in various projects for neural network design, realization and testing. His activity is proved by more than a hundred scientific contributions and books.

Fausto Sargeni was born in Riano (ROMA) in 1961. He received the Dipl. Eng. degree in Electronic Engineering from the University of Rome "La Sapienza" in 1987. In 1989 he joined the Department of Electronic Engineering, University of Rome "Tor Vergata" as assistant professor. His research interests include analog VLSI circuits for Neural Networks and nonlinear circuits.

Vincenzo Bonaiuto was born in Rome, Italy, in 1962. He received the Electronic Engineering Degree at the University of Rome "La Sapienza" in July 1992. In 1993 he received the Diploma of specialization in Telecommunication at the ISPT in Rome. At moment he finished his PhD courses in Telecommunication and Microelectronics and is assistant professor at the University of Rome "Tor Vergata."

Analog Integrated Circuits and Signal Processing, 15, 251–262 (1998)

Analog VLSI Design Constraints of Programmable Cellular Neural Networks

PETER KINGET[1] AND MICHIEL STEYAERT[2]

[1]*ESAT-MICAS Laboratory, Katholieke Universiteit Leuven, 3001 Leuven, Belgium; Current address: Bell Labs, Lucent Technologies, Murray Hill, NJ 07974-0636 USA;* [2]*ESAT-MICAS laboratory, Katholieke Universiteit Leuven, Kardinaal Mercierlaan 94, 3001 Leuven, Belgium*

Received August 8, 1996; Accepted January 27, 1997

Abstract. Analog parallel signal processing systems, like cellular neural networks (CNN's), intrinsically have a high potential for perception-like signal processing tasks. The robust design of analog VLSI requires a good understanding of the capabilities as well as the limitations of analog signal processing. Implementation-oriented theoretical methods are described to compute the effect of all types circuit non-idealities with random or systematic causes on the static and dynamical behavior of CNN's and to derive specifications for the cell circuit building blocks. The fundamental impact of transistor mismatch on the trade-off between the speed, accuracy and power performance of CNN chips is demonstrated. A design methodology taking into account the effect of transistor mismatch is proposed and experimental results of a CNN chip implementation designed with this method are discussed.

Key Words: cellular neural networks, transistor mismatch, analog signal processing, analog VLSI, analog circuit design, MOS circuit design

I Introduction

Massively parallel signal processing systems are expected to open new possibilities for the solution of challenging signal processing problems in perception applications. Several realizations have demonstrated or explored the potentials of analog VLSI technologies for the realization of these parallel systems. Analog circuits indeed offer many advantages for the implementation of these systems but it is very hard to attain the same level of robustness as with digital circuits. It is therefore of prime importance that the capabilities as well as the limitations of analog integrated circuits are well understood in order to obtain high performance signal processing systems.

In this paper a design methodology for the robust design of programmable cellular neural networks (CNN) [1] is described. The analog signal processing circuits of a programmable CNN form the core of the analog circuitry of a cell of a CNN universal machine (CNNUM) [2] and consequently the results and methodologies presented in this paper can be directly applied for the design of the analog CNNUM hardware. To start a circuit design, an analog circuit designer needs specifications for the different building blocks of the system that allow him to understand the requirements and characteristics of the signal processing system. In section II we develop specifications and limits for the different circuit non-idealities. Good circuit specifications are crucial since an over-specification of the circuit requirements will result in poor system performance whereas bad specifications will result in malfunctioning realizations. As such, the translation of the requirements of the system for a correct operation into circuit specifications inherently fixes the attainable performance of the circuits.

For the circuit realization of a CNN, area compactness is one of the primary goals and the size of the components is reduced as much as possible. Unfortunately the statistical variation in the components, the transistor mismatch and the signal processing inaccuracy increase when the size is reduced. In section III the fundamental impact of transistor mismatch on the ultimate performance of CNN is discussed.

In section IV a design methodology for programmable CNN's is developed that accounts for the effects of transistor mismatch in order to obtain a robust circuit design. Experimental results of a fully programmable analog implementation of a 20 × 20 CNN, designed using this methodology are presented.

Notations and Terminology In figure 1 a principle schematic of an electrical implementation of a CNN cell is represented; the following notations are used:

x_i: state of cell i
u_i: input of cell i

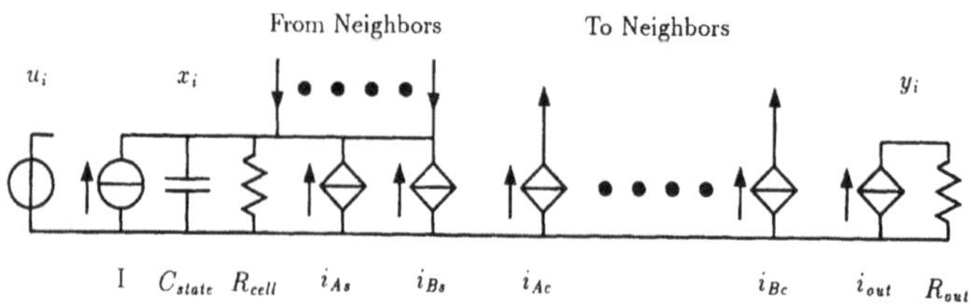

Fig. 1. Schematic for an electrical CNN cell implementation: $i_{Ac} = A_c \cdot y_i$, $i_{Bc} = B_c \cdot u_i$, and $i_{out} = 1/2R_{out}(|x_i + 1| - |x_i - 1|)$.

y_i:	output of cell i; $y_i = f(x_i)$
$f(x_i)$:	PWL output non-linearity; $y_i = f(x_i) = \frac{1}{2}(\lvert x_i + 1\rvert - \lvert x_i - 1\rvert)$
R_{cell}:	the cell state resistor
C_{state}:	the cell state capacitor
$\tau = R_{cell} \cdot C_{state}$:	the time constant of the state

II From System Description to Circuit Specifications

The worlds of system theory and analog circuit use different conventions, have different resources and different constraints. In theoretical definitions of systems a much higher level of abstraction can be used. However, practical analog circuit realizations of computational circuits have many unwanted properties and imperfections compared to their ideal description. To make a successful circuit implementation of a new signal processing paradigm the circuit designer needs to know the sensitivity of the system to the imperfections of analog circuits. This knowledge is typically presented in the form of circuit specifications. In this area only limited information is available from the system-level definitions; to bridge this gap extra implementation-oriented theory is necessary that specifically deals with the implications of circuit imperfections on the system operation.

The VLSI imperfections can have a systematic or a random cause and can have a detrimental effect on the static or the dynamical circuit behavior so that basically 4 types of errors can be distinguished: random static, random dynamic, systematic static, and systematic dynamic errors. The different causes of the errors as well as their effect on the different circuit specifications are discussed in the subsequent sections.

II.A Random Static Errors and Accuracy Specifications

The stochastic processes associated with the different fabrication steps of integrated circuits result in a statistical variation of the properties of transistors and consequently in a mismatch between the characteristics of geometrical identical devices [3], [4]. These mismatches are time-invariant and mainly result in a static random error in the template weight values; their effect on the cell dynamics is discussed in II.B.

A large random variation of the weight values from cell to cell can result in an incorrect cell behavior [5], [6]. The accuracy specifications for the template weights describe the allowed random variation in the weight values that still guarantee a correct CNN operation. In order to obtain accuracy specifications the robustness of CNN templates for random static error must be investigated [7]: first the conditions for a correct operation of a single cell are derived; then the conditions for a correctly operating network are established and the accuracy specifications that guarantee a correct network operation with a given yield are calculated. In figure 2 the accuracy requirements for different templates to obtain a 95% yield for a 400 cell (20×20) network are shown.

The robustness evaluation method also provides insight in the causes of high or low robustness of specific templates. This information can be used for the optimization of templates for high implementability [7], [8]. The availability of robust templates is essential for the design of high performant CNN chips; template robustness is one of the most clear demonstrators of the importance of a good system definition and optimization towards hardware implementation.

In section IV the derived accuracy specifications are used for the minimization of the transistor areas to obtain a high cell density but with a guarantee of a correct network operation.

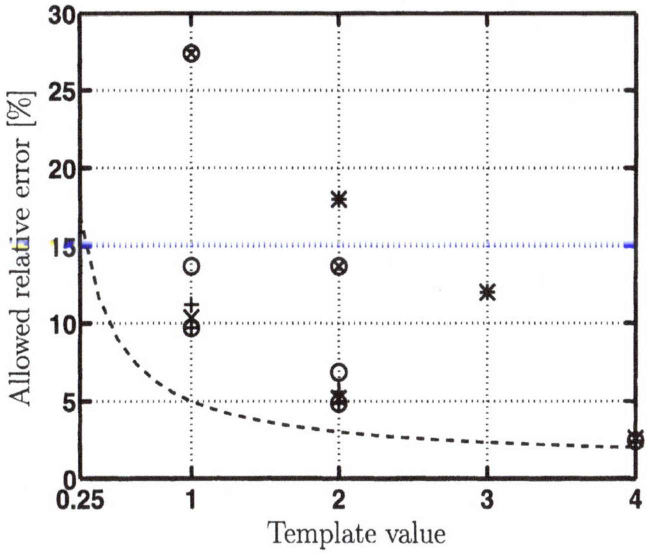

Fig. 2. The allowed relative error for the template values for a 20x20 network and a yield of 95%: o = CoCoD, x = HOLE, ⊗ = SHADOW, ⊕ = EDGE, + = NOISE, and * = PEEL templates [9]. The specification for the programmable template circuit is (- -).

II.B Random Dynamical Errors

Random differences in the dynamical behavior of different cells in a CNN originate from the random variation of the template weights and the time constants of the integrator in the cell, which depends on the cell's state capacitance C_{state} and the cell resistor R_{cell} in an analog implementation. These variations in the cell dynamics can cause an incorrect interaction between the cells and result in an erroneous computation and a random dynamical error. Their impact is different for the different types of templates.

For a *non-propagating template* [7]—a template with an A template equal to zero except for the self-feedback—all cells evolve independently so that no dynamical errors can occur. A CNN programmed with *a propagating template*—a template with an A template non-equal zero—is a very complex dynamical system with non-linear feedback-loops within and over the cells and one can expect incorrect behavior if variations in the dynamics exist.

However, the correct operation of many *propagating templates with input*— templates with a non-zero A and a non-zero B template—is insensitive to the dynamics of the different cells. In a holefiller [9] e.g. the original input is stored on the cell's input and all cells are initialized to black; during the evolution, white waves evolve from the edges towards the center of the image. If all cells operate correctly statically, a cell only becomes white if it is indeed not part of a hole. Even if a cell evolves slowly compared to the other cells, as long as it operates correctly and only becomes white if necessary, its slower transition can then still propagate through the network and no errors occur. Once a cell has evolved towards white from its black initial state, there is no way back so that an incorrect decision to evolve to white must be avoided, but the correct static operation indeed guarantees that such errors do not occur. The shadow generator template and many other propagating templates with input basically operate in a similar way and the same argument can be used to conclude that they are insensitive to dynamical variations. The yield of the networks programmed with these templates is then only dependent on the correct static operation of the cells discussed in II.A.

Propagating templates without input—templates with a non-zero A and a zero B template—use the input image as the initial state and during their computation, this original input is not stored in the network. Consequently this class of templates inherently has a much higher susceptibility to dynamic errors. The connected component detector and the averaging template are examples of such templates [9]. However, Monte-Carlo simulations accounting for all implications of the random variations show that the static effect of template variations is much more significant than their dynamical effect [7], [8], so that reliable accuracy specifications can be derived as outlined in section II.A.

We can summarize that the effect of random dynamical errors on the correct operation of CNN's is very small. Non-propagating templates and propagating templates with input are insensitive to variations in the dynamics and for propagating templates without input, the effect of random errors on the correct static behavior is more significant than the effect on the dynamical behavior.

II.C Systematic Static Errors and the Application of Tuning Techniques

Systematic errors are caused by deterministic processes and are the same for every cell. The non-linearity of device characteristics can result in a signal-dependent behavior of the cell circuit blocks and can introduce distortion errors. These errors can be eliminated by correct biasing, good signal amplitude choices, or non-linearity cancellation schemes like differential or trans-linear circuit techniques. Random and systematic static errors differ in nature of their cause but their impact on the network behavior can be evaluated with a similar methodology. We can thus evaluate the effect of a systematic static error by controlling the correct operation of a single cell from its dynamic routes.

In advanced digital CMOS processes no linear resistors with a high resistance are available, and an active circuit comprising of only transistors must be used to emulate the *cell state resistor* [10]. In figure 3(a) the characteristic of such an implementation is shown and in figure 3(b) the effect on the shape of a dynamic route[1] of a connected component detector cell with two white or two black neighbors is illustrated. The resistor distortion does not introduce significant differences in the cell dynamic routes as long as its behavior in the unit state ($x_i = 1$ or $x_i = -1$) is matched to the ideal behavior. The effect of the imperfections of the circuit implementation of the *PWL output non-linearity* can be evaluated using the same dynamic route technique [8] and as long as a sigmoid type of non-linearity is used a correct CNN behavior is guaranteed [1].

Non-linearity errors in the *A template multipliers* for a programmable CNN implementation result in a reduction of the effective weight value with the output value and change the slope of the output non-linearity; as discussed above this has no impact on the correct behavior. However, care has to be taken that the output signal in both saturation regions ($x_i \leq -1$ or $x_i \geq 1$) is correct. By using a tuning loop to derive the control signals for the template multipliers this effect can be compensated. Due to the distortion in the *B-template multipliers*, the feed-forward signals are not linear scalings of the input of the cell. Applications using gray-scaling input images e.g. will suffer from distortion in the signal processing. For applications with binary inputs a similar tuning strategy as for the A-template multipliers can be applied.

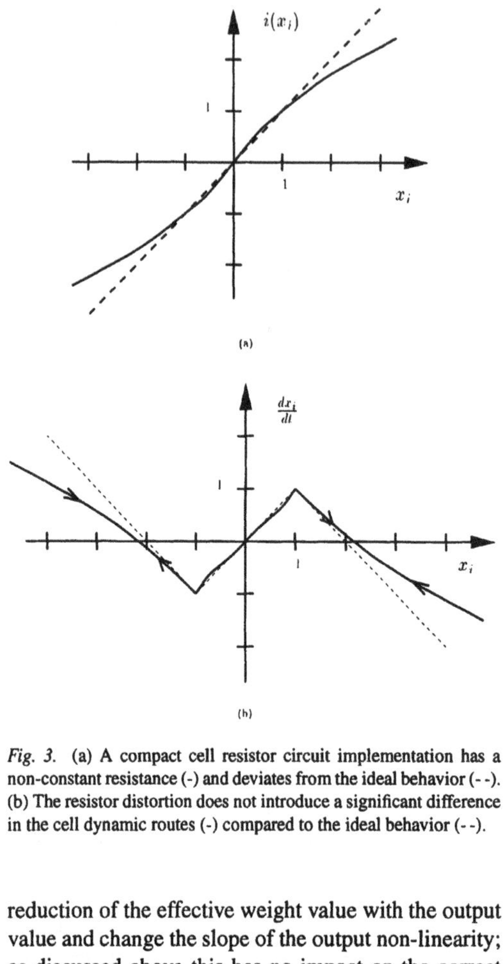

Fig. 3. (a) A compact cell resistor circuit implementation has a non-constant resistance (-) and deviates from the ideal behavior (- -). (b) The resistor distortion does not introduce a significant difference in the cell dynamic routes (-) compared to the ideal behavior (- -).

We summarize that the effect of the systematic static errors can be eliminated by an on-chip automatic bias and weight tuning system. Since the distortion errors are systematic errors, a single bias circuit and tuning loop for the whole network is sufficient, so that the area

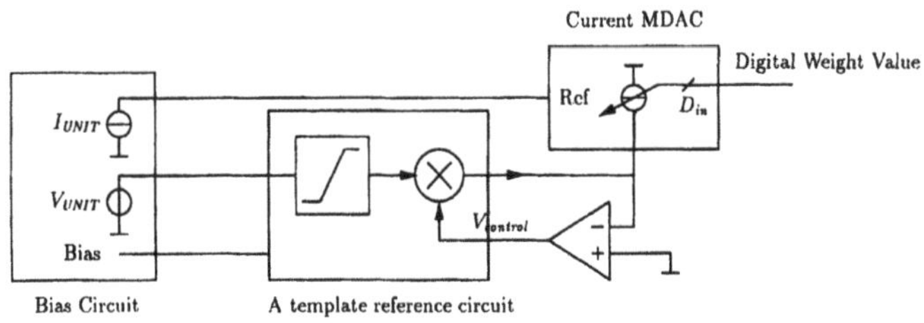

Fig. 4. Block diagram of an on-chip bias and weight tuning system for transconductance type template multipliers.

penalty is very small. In figure 4 a block diagram of a tuning system for transconductance type cell multipliers is shown. The bias circuit uses the cell resistance as a reference to bias the output non-linearity function circuit correctly and to derive the value of the on-chip unit voltage reference V_{UNIT}, which represents $x_i = 1$ in voltage, and the the value of the on-chip unit current voltage reference I_{UNIT}, which represent $x_i = 1$ in current [10]. These references are then used to compute the control signals ($V_{control}$) for the A template multipliers with a tuning loop that adjusts the control signal so that the output signal reaches the correct level for the desired template weight [10]. The correct output current signal is computed with a multiplying digital to analog converter, using the on-chip generated reference current, from a digital, user-specified template value.

II.D Systematic Dynamical Errors

Every node in a circuit has a (parasitic) capacitance and is driven by a signal source with a limited conductance, so that the node voltage cannot change instantaneously but will always require a certain time constant. Most of these time constants are much smaller than the dominant cell time constant so that they have very little impact. Unfortunately, in all practical circuits there is always more than one time constant that has a considerable effect on the operation. In a CNN cell, a dominant time constant is deliberately introduced at the state node by the effect of the state capacitor and the state resistor. The output node in a CNN must drive the inputs of all the A-template multipliers (see fig. 1) and in a circuit realization of a programmable CNN, the capacitive loading of the output node is large. Consequently, it is a very realistic and general assumption that the second time constant in the cell circuit—and thus a parasitic integrator—is located at the output node. This second time constant introduces extra delays in the feedback loops created by the feedback template A; a loop exists in the cell due to the self-feedback, and feedback loops, that span over several cells, also exist for propagating templates. For a circuit designer, it is important to know how large the parasitic time constants can be relatively to the cell time constant, so that the circuit still functions correctly and remains stable.

In a CNN programmed with *a non-propagating template*, there is only a feedback loop within every cell, due to the necessary self-feedback to obtain a binary output. To investigate the effect of the second time constant on the stability of the cells, we cannot use the classical linear feedback theory, due to the non-linear output function f. The behavior of the cell with two time constants is modeled with the following state space equations:

$$\begin{aligned} \tau_s \frac{dx_i}{dt} &= -x_i + A_i y_i + K \qquad (1) \\ \tau_o \frac{dy_i}{dt} &= -y_i + f(x_i) \end{aligned}$$

where x_i and y_i are the state and output of the cell, τ_s and τ_o are the time constants of respectively the state and the output, A_i is the self-feedback weight in the A template, $f(x_i)$ is the PWL output nonlinearity and K is the sum of the feed-forward signals. The equilibria of the cell are the pairs $\{x_i, y_i\}$ for which the left hand sides of (1) are equal to zero and for which the time derivatives are thus zero.

We must now investigate if the cell behavior is different due to the extra integrator that is present at the output of the cell. In most non-propagating templates, A_i is equal to 2 [9]. The most critical situation, then occurs when K is 0. When no second time constant is present ($\tau_o = 0$), the cell state x_i has two stable equilibria at -2 and at 2; for an initial state smaller than 0

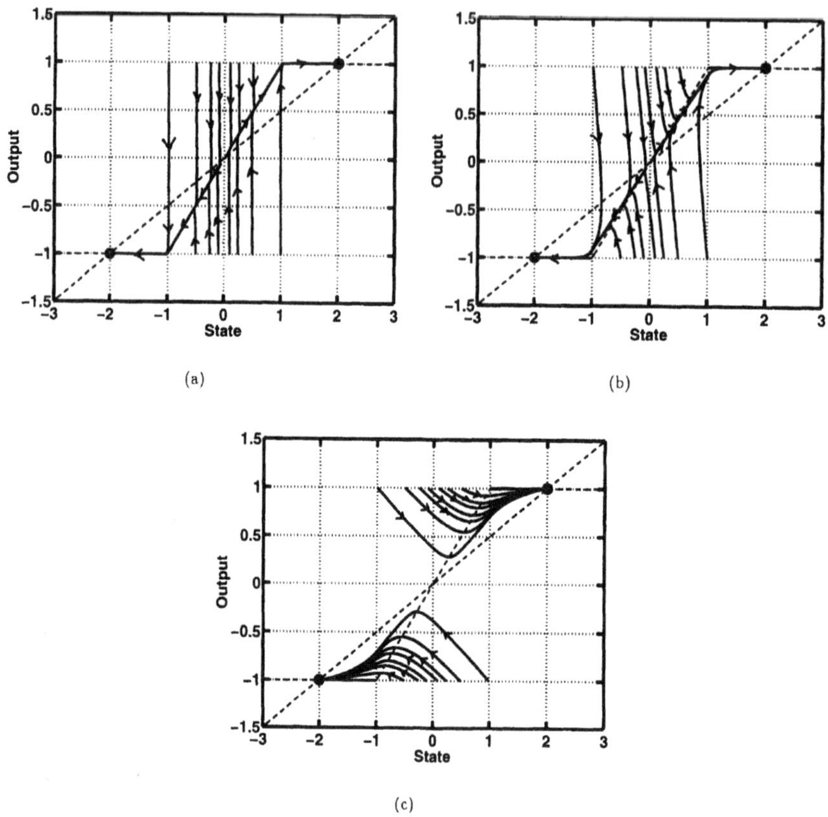

Fig. 5. The state space trajectories (-) for the state x_i and the output y_i for different initial situations and different ratio's of the time constants: (a) $\tau_s/\tau_o = 100$, (b) $\tau_s/\tau_o = 10$, and (c) $\tau_s/\tau_o = 1$. The stable equilibria are indicate by o; they are at the crossing of the (- -) curves which represent the model equations in (1) for zero time derivatives.

the cell state x_i evolves to -2 and for an initial state larger than 0 the cell state evolves to 2; at 0 an unstable equilibrium exists.

In order to study the effect of the second time constant, we draw the state-space trajectories for the two-time constant system in (1). We suppose the cell can only start with an initial value for the output y_i of -1 or 1, which is the output of the previous computation. The state-space trajectories for different initial states and different ratios τ_s/τ_o are drawn in figure 5:

- for a τ_o which is 100 times smaller than τ_s, the behavior of the cell is almost ideal (figure 5(a)); the cell evolves to a final state of 2 in the initial state is larger than 0 and to a final state of -2 if the initial state is smaller than 0, *independent* of the initial value of the output node.
- For a τ_o 10 times smaller than τ_s, we see that for very small initial states, larger than 0, and for an initial value of the output of -1, the cell evolves to the equilibrium state at -2, which is different from the ideal behavior. For larger initial states the behavior is still correct (figure 5(b)).
- For a τ_o equal to τ_s, the cell state evolves to the equilibrium at -2 for all positive initial states if the output starting value is -1. At this point the behavior of the cell is incorrect for all positive initial states (figure 5(c))! Similarly an incorrect behavior occurs for the negative initial states.

The second time constant indeed has an influence on the behavior of a cell in a CNN with a non-propagating template. Not only the initial state and the inputs of the cell and its neighbors, which is modeled by K in

(1), but also the previous value of the output of the cell determine its final state. As a rule of thumb, the parasitic τ_o should be made at least 10 times smaller than the wanted cell time constant τ_s and that the cell behavior is then very close to the ideal behavior.

In a CNN programmed with *a propagating template*, feedback loops exist that span over two cells. If we include the parasitic integrator at the output of the cell, we obtain a feedback loop with 4 poles and 2 non-linear functions f and very complex dynamic behavior is now possible. The problem of the stability of this system is somewhat related to the stability of delay-type template CNN's [11], but there a fixed delay is used and not a first-order time constant. A detailed study of the stability of these high order non-linear feed-back loops is very involved and is still a topic of intensive research. However, if the parasitic time constant at the output is sufficiently small, the cell will almost behave as in the ideal case. Therefore, as a rule of thumb we will require to make the parasitic time constant 10 times smaller as the cell time constant. This approach is similar to the stabilization technique that is used in many classical linear feedback circuits with high order poles [12]. Instability occurs if the poles approach each other, and a straightforward way of stabilizing the loop is by adding extra capacitance to make one time-constant considerably larger than the others. Then the loop behaves almost as a one pole system and is stable.

The specification of the relative position of the two time constants, has important implications for the minimal power consumption of the circuits. To reduce the unwanted time constant to a 10x smaller value as the dominant state time constant, the impedance level in the circuit must be reduced; this implies that the bias current in the transistors must be increased to increase their (trans)conductance so that the power consumption of the circuit increases. It is clear that the better the specification of the relative position, the better performing circuits are obtained.

II.E Summary

The effect of the different types of implementation imperfections has been treated. If on-chip automatic biasing and tuning techniques are applied the effect of systematic static errors can be eliminated very effectively and their impact on the CNN performance is small. The most important systematic dynamical error is the second time constant at the cell's output which must be made 10 times smaller as the cell time constant; this has important implications on the speed or power consumption of CNN chip implementations.

The effect of random errors on the dynamical behavior of CNN cells is negligible compared to their effect on the static behavior. Using statistical techniques the implications of random static errors on the yield of CNN chips is calculated and accuracy specifications for the the template weight circuits are determined. In the next section we will show that these random static errors caused by mismatches in transistors have very important consequences on the circuit sizing and impose a fundamental limitation on the ultimate performance of analog parallel signal processing systems.

III Implications of Transistor Mismatch

The parameters of two geometrically identical MOS transistors are not completely identical and their difference has a normal distribution with a zero mean and a standard deviation that depends on the geometry of the devices [3], [4]. The difference of the threshold voltage ΔVT has important implications for the total performance of analog circuits [13]; this difference is modeled by:

$$\sigma^2(\Delta V_{TO}) = \frac{{A_{VTO}}^2}{W\,L} \tag{2}$$

where W and L are respectively the width and the length of the transistor and A_{VTO} is a technology matching constant.

To achieve a given signal processing accuracy in a circuit, a given matching of the devices, or $\sigma(\Delta V_{TO})$, is necessary and consequently a minimal $\sqrt{\text{area}}$ or $\sqrt{W\,L}$ for the devices is imposed. In CMOS design this implies that the only free design parameter is the $\left(\frac{W}{L}\right)$ of the devices. By using small L, for instance, the intrinsic speed of the transistors is large and high speed circuits are obtained; however, to meet the accuracy specification a large W and consequently a large current and power consumption is implied. As a result, the effect of transistor mismatch on the circuit design imposes a fundamental limit to the total performance of a circuit and the optimal trade-off between specifications given by the following expression [13], [8]:

$$\frac{\text{Speed}\cdot\text{Accuracy}^2}{\text{Power}} \propto \frac{1}{C_{ox}{A_{VTO}}^2} \tag{3}$$

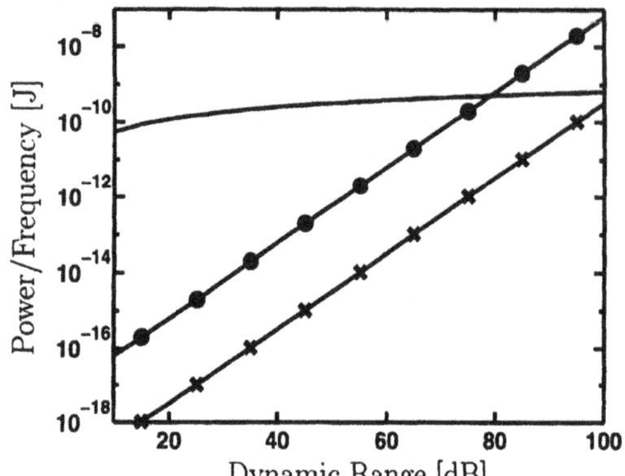

Fig. 6. The energy consumption for a digital implementation (–) of a pole as a function of the required dynamic range; the minimal energy consumption of analog implementations due to the impact of mismatch (–o–) or noise (–x–).

where the proportionality constant is fixed by the circuit architecture; C_{ox} is the transistor gate capacitance per unit area and is a technology constant inversely proportional to the gate-oxide thickness. Equation (3) explicitly states that the different performances can only be exchanged one for the other, and that the total performance[2] *Speed·Accuracy²/Power* is fixed by the used VLSI technology.

Moreover mismatch imposes a boundary on the minimal power consumption P of an analog VLSI signal processing system to achieve a given accuracy or dynamic range DR and a given speed or signal frequency f [13]:

$$P = 24\, C_{ox}\, A_{VTO}{}^2\, f\, DR^2 \tag{4}$$

which is valid for any analog circuit rely-ing on component matching to achieve its accuracy. These mismatch implications have very important consequences for the maximal performance of analog parallel VLSI signal processing systems and CNN chips.

In a (programmable) CNN chip which can execute one or several templates, the necessary accuracy is fixed by the templates, the wanted yield, and the network size. The Power/Speed ratio of the chip is then fixed for a given technology and depends quadratically on the necessary accuracy. Good accuracy specifications are thus crucial to a high performance CNN chip design and at the same time template robustness should be maximized as much as possible.

Thermal noise also imposes a limit of the power consumption P of analog circuits for a given operation frequency f and dynamic range DR, expressed as: $8\, kT\, f\, DR^2$ [14], where k is the Boltzmann constant and T the absolute temperature. In figure 6 this noise limit and the mismatch limit of equation (4) on the Power/Speed ratio or the energy consumption per cycle are plotted as function of the dynamic range DR or accuracy; also the energy per cycle for a typical digital implementation of a single pole transfer function is plotted after [14]. The mismatch limit is about two orders of magnitude more important than the thermal noise limit so that the circuit design in CNN chips will be limited by mismatch implications. Figure 6 also illustrates that an analog implementation is advantageous over a digital implementation for systems requiring only a low accuracy for the computations. The steep slope of the energy consumption of an analog implementation as function of the dynamic range or accuracy again illustrates the importance of good accuracy specifications.

Since mismatch is a fundamental limitation to the performance optimization of CNN chips, the scaling of the matching behavior as function of the technology feature size—commonly denoted as λ—is an important question. The matching quality of a technology is inversely proportional to the $C_{ox} A_{VTO}{}^2$. C_{ox} is proportional to the gate-oxide thickness which is on its turn proportional to the technology size λ. Mismatches in

Given: Mismatch models technology
Device models technology
Circuit & accuracy expressions

Constraints: ACCURACY specifications (Yield)
SPEED specification
Correct Biasing and Circuit Operation

⇒**Optimize** the signal swings
the bias points
the device sizes

⇒ to **Minimize** CELL AREA

Fig. 7. Outline of the constrained optimization problem for sizing the cell circuits; the cell density is maximized while guantee-ing a correct cell operation by controlling the cell accuracy in every step through the accuracy constraints.

Table 1. First order estimation of the scaling of the performances of an analog implementation of a programmable CNN [15] as function of the technology feature size; the required accuracy is assumed constant and a constant speed and constant power scaling scenario is illustrated for a technology with a two-times smaller feature size.

		Constant POWER		Constant SPEED	
	0.7 μm CMOS	$\alpha = \frac{0.7}{0.35}$	0.35 μm CMOS	$\alpha = \frac{0.7}{0.35}$	0.35 μm CMOS
ACCURACY	see fig. 2	$\equiv$	see fig. 2	$\equiv$	see fig. 2
SPEED ($\propto$ Cell time constant)	$1/(4.8\ \mu s)$	$\times\,\alpha$	$1/(2.4\ \mu s)$	$\equiv$	$1/(4.8\ \mu s)$
POWER Con.Comp.Det.	375 μW/cell	$\equiv$	375 μW/cell	$\times\frac{1}{\alpha}$	187.5 μW/cell
DENSITY ($\propto$ 1/Area)	16.7 cells/mm^2	$\times\,\alpha^2$	66.8 cells/mm^2	$\times\,\alpha^2$	66.8 cells/mm^2

Fig. 8. Micro-photograph of the fully programmable 20 × 20 CNN chip containing 400 cells, I/O, template weight tuning and control circuits.

the depletion charges under the gate are the dominant physical cause for V_{T0} mismatch; this charge mismatch is transformed in a voltage mismatch by the gate capacitance and consequently A_{VT0} is proportional to the gate-oxide thickness and the technology size λ [3]. Consequently the *Speed·Accuracy²/Power* scales inversely proportional with the technology size λ. The accuracy is inversely proportional to $\sigma(\Delta V_{T0})$, and from equation (2) we then derive that the area of the circuit is proportional to square of the wanted accuracy and the $A_{VT0}{}^2$ of the technology: Area $\propto$ Accuracy$^2 \cdot A_{VT0}{}^2$; if the accuracy requirements remain constant the area of

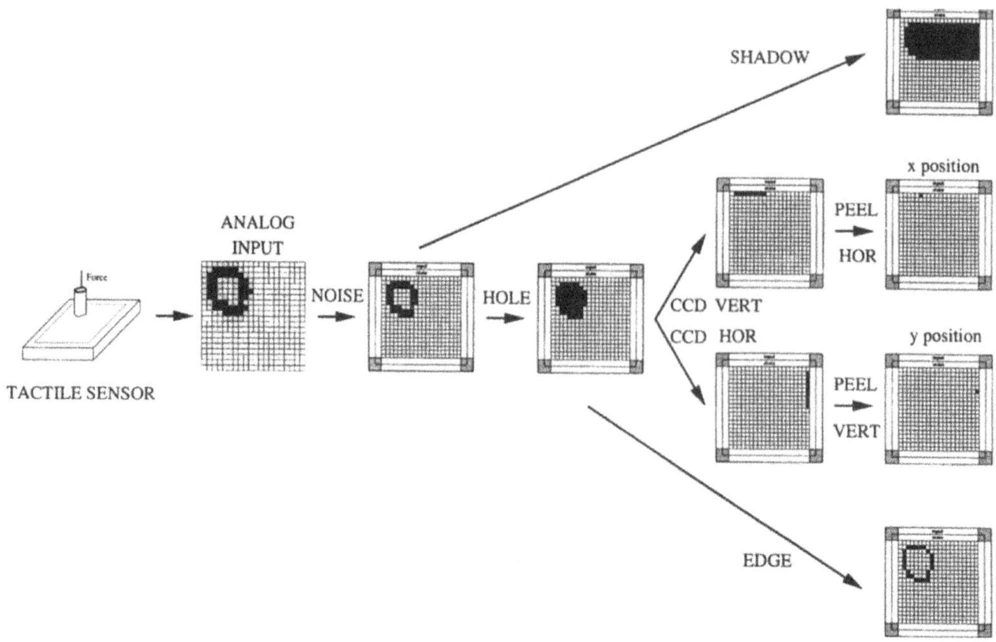

Fig. 9. Measured results of template sequences (from table [9]) executed on the tactile sensor output.

a circuit scales quadratically with the technology size λ. Using these scaling laws, a first order estimation for the scaling of the performance of an analog programmable CNN chip implementation is illustrated in table 1. These scaling laws do not take into account the effect of the reduction of the power supply voltage and consequently of the signal levels for smaller feature size technologies; this signal level reductions can however degrade the performance so that the beneficial effect of scaling is reduced. For very deep-sub-micron CMOS technologies ($\lambda < 0.2\ \mu m$), the mismatching of other transistor parameters than V_{T0} also becomes significant and the scaling laws have to be modified [8].

IV Design Methodology for Programmable CNN's

The methods and results presented in section II allow to clearly specify the requirements of the circuit building blocks of a cell implementation; these specifications are the foundation to start the circuit design and they guide the selection of the different cell circuit implementations.

Once a satisfactory circuit schematic has been developed the sizing of the devices has to be performed. As outlined above, the design of a CNN chip is dominated by the effect of transistor mismatch and consequently a design methodology must be applied that accounts for transistor mismatch. Three critical parameters exist for a CNN chip: speed, power consumption and accuracy, which is directly related to area or density. However, due to the impact of mismatch on the circuit design, only two parameters can be fixed and the third is implied by the technology quality (see equation (3)). The required accuracy of the circuits is derived from the template set, the size of the network and the wanted (high) parametric yield for the chips (see figure 2). Since the accuracy is fixed, and once the speed of the cells is specified, in first order the power consumption is implied by technology. However, in large analog circuits several degrees of freedom exist in the design that still can be optimized and this optimization corresponds to the minimization of the proportionality constant in equation (3), which is determined by the circuit architecture.

The resulting optimization problem is a constrained optimization problem and is outlined in figure 7. Mismatch models are essential technology information, together with the classical device models. The behavior of the circuits must be expressed analytically. The accuracy and speed specifications are part of the constraint set and are complemented with constraints that guarantee a correct biasing and operation of the analog circuitry. These constraints fix a number of design variables such as aspect ratios $(\frac{W}{L})$ or areas of devices, biasing points or signal swings, but typically there remain a number of free design parameters. These are then optimized towards a minimal area for instance. This constrained sizing optimization problem of the total cell circuit is not straightforward and cannot be solved analytically. The sizing model of the circuit and the constraints are programmed in a general purpose mathematical tool or a specific tool for analog circuit design and an optimum is computed. The correct operation and high yield of the circuits is guaranteed thanks to the accuracy specifications which are satisfied in every optimization step.

This design procedure has been applied for the design of a fully programmable analog implementation of a 20×20 CNN, using the Chua-Yang model [15]. A chip photograph is shown in figure 8. The measured performance of this chip, fabricated in a 0.7 μm digital CMOS technology, is summarized in the second column of table 1. All cells contain 5 A template multipliers (north, south, east, west, feedback connections) and 5 B template multipliers and the weight values can be programmed over a continuous range between $\pm 1/4$ and ± 4. Template weight tuning circuits are included on-chip to eliminate systematic static errors; their occupy very little area and are located between the bonding pads. The function of the chip is controlled by a personal computer and an efficient X-Y addressing Input/Output block interfaces the chip to a two-dimensional sensor [16]. The chip is used as a real-time sensor signal processing system and has been tested with a tactile sensor; in figure 9 the experimental results of template sequences—the templates are taken from [9]—executed on the tactile sensor output are presented.

V Conclusions

The design of high performance integrated analog signal processing systems requires a good understanding of the trade-off between the system or application requirements and the capabilities and limitations of analog integrated circuits. Only if these trade-offs are fully understood, high performance analog signal processing systems can be developed that can compete in specific application domains with the widespread digital signal processing systems.

Implementation-oriented theory has been developed to calculate the effect of random or systematic circuit non-idealities on the static and dynamic behavior of CNN chips. Specifications for the different types of errors have also been derived. This full set of specifications form the foundation for a robust analog CNN chip design. We have demonstrated that the trade-off between the speed, accuracy and power performance of analog CNN implementations is fundamentally limited by the impact of transistor mismatch and the influence of technology downscaling on the CNN chip performance has been discussed. Moreover, a chip design methodology to account for transistor mismatches during the circuit sizing has been proposed and experimental results for a fully programmable 20×20 analog CNN chip implementation, designed with this methodology, have been discussed.

Notes

1. This is a very typical example of a critical dynamic route for any template with a self-feedback of 2 in a situation where the net effect of the neighbor actions is 0 (see e.g. [9]).
2. *Speed·Accuracy²/Power* expresses the total quality of the circuit design, since a high speed, high accuracy and low power consumption are desirable.

References

1. L. O. Chua and L. Yang, "Cellular neural networks: Theory." *IEEE Transactions on Circuits and Systems* 35(10), pp. 1257–1272, 1988.
2. T. Roska and L. O. Chua, "The CNN universal machine: an analogic array computer." *IEEE Transactions on Circuits and Systems—II: Analog and Digital Signal Processing* 40(3), pp. 163–173, March 1993.
3. M. Pelgrom, A. Duinmaijer, and A. Welbers, "Matching properties of MOS transistors." *IEEE Journal of Solid-State Circuits* 24(5), pp. 1433–1439, 1989.
4. J. Bastos, M. Steyaert, R. Roovers, P. Kinget, W. Sansen, B. Graindourze, N. Pergoot, and E. Janssens, "Mismatch characterisation of small size MOS transistors," in *Proceedings of the IEEE International Conference on Microelectronic Test Structures*, March 1995, pp. 271–276.
5. B. E. Shi, S. Wendsche, T. Roska, and L. O. Chua, "Random variations in CNN templates: Theoretical models and empirical studies," in *Proceedings International Workshop on Cellular Neural Networks and their Applications (CNNA)*, December 1994, pp. 27–32.
6. P. Kinget and M. Steyaert, "Evaluation of CNN template robustness towards VLSI implementation," in *Proceedings International Workshop on Cellular Neural Networks and their Applications (CNNA)*, December 1994, pp. 381–386.

7. P. Kinget and M. Steyaert, "Evaluation of CNN template robustness towards VLSI implementation." *International Journal of Circuit Theory and Applications* 24, pp. 111–120, Jan. 1996.
8. P. Kinget and M. Steyaert, *Analog VLSI integration of massive parallel signal processing systems*, Kluwer Academic Publisher: Norwell, MA, ISBN 0-7923-9823-8, 1997.
9. T. Roska and L. Kek (Editors), "Analogic CNN program library," Analogical and Neural Computing Laboratory, Computer and Automation Institute of the Hungarian Academy of Sciences, Budapest, Hungary, Tech. Rep. DNS-5-1994, June 1994.
10. P. Kinget and M. Steyaert, "A programmable analog cellular neural network CMOS chip for high speed image processing." *IEEE Journal of Solid-State Circuits* 30(3), pp. 235–243, March 1995.
11. T. Roska and L. O. Chua, "Cellular neural networks with non-linear and delay-type template elements and non-uniform grids." *International Journal of Circuit Theory and Applications* 20, pp. 469–481, 1992.
12. K. R. Laker and W. M. C. Sansen, *Design of Analog Integrated Circuits and Systems*, McGraw-Hill, 1994.
13. P. Kinget and M. Steyaert, "Impact of transistor mismatch on the speed-accuracy-power trade-off of analog CMOS circuits," in *Proceedings of the IEEE Custom Integrated Circuits Conference (CICC)*, May 1996, pp. 333–336.
14. E. A. Vittoz, "Future of analog in the VLSI environment," in *Proceedings ISCAS*, May 1990, pp. 1372–1375.
15. P. Kinget and M. Steyaert, "An analog parallel array processor for real-time sensor signal processing," in *Digest of Technical Papers IEEE International Solid-State Circuits Conference (ISSCC)*, Feb. 1996, pp. 92–93.
16. P. Kinget and M. Steyaert, "Input/output hardware strategies for cellular neural networks," in *Proceedings of IEEE International Conference on Neural Networks (Orlando)*, July 1994, pp. 1899–1902.

Peter R. Kinget was born in 1967. He received a masters degree in electrical and mechanical engineering and a Ph.D. degree in electrical engineering from the Katholieke Universiteit Leuven (Belgium) in 1990 and 1996 respectively.

From 1991 to 1995 he received a fellowship from the Belgian National Fund for Scientific Research (NFWO) that allowed him to work as a research assistant at the ESAT-MICAS laboratory of the Katholieke Universiteit Leuven. In October 1996 he joined Bell Laboratories, Lucent Technologies in Murray Hill, NJ as a Member of Technical Staff.

His research interests are in analog telecommunication circuits, transistor mismatch and analog parallel signal processing systems.

Prof. M. Steyaert received his Ph.D. degree in electronics from the Katholieke Universiteit Leuven, Belgium in 1987.

In 1988 he was a Visiting Assistant Professor at the University of California, Los Angeles. Since 1989 he has been appointed by the NFWO at the Laboratory ESAT, K.U Leuven, where he has been an Associate Professor since 1990 and Research Director since 1996.

His current research interests are in high-frequency analog integrated circuits for telecommunications and analog signal processing.

Analog Integrated Circuits and Signal Processing, 15, 263–275 (1998)

Focal-Plane and Multiple Chip VLSI Approaches to CNNs

M. ANGUITA, F. J. PELAYO, E. ROS, D. PALOMAR AND A. PRIETO

Departamento de Electrónica y Tecnología de Computadores, Facultad de Ciencias, Universidad de Granada, 18071-Granada, Spain

fpelayo@ugr.es

Received October 1, 1996; Accepted November 26, 1996

Abstract. In this paper, three alternative VLSI analog implementations of CNNs are described, which have been devised to perform image processing and vision tasks: a programmable low-power CNN with embedded photo-sensors, a compact fixed-template CNN based on unipolar current-mode signals, and basic CMOS circuits to implement an extended CNN model using spikes. The first two VLSI approaches are intended for focal-plane image processing applications. The third one allows, since its dynamics is defined by process-independent local ratios and its input/outputs can be efficiently multiplexed in time, the construction of very large multiple chip CNNs for more complex vision tasks.

1. Introduction

The **CNN** model introduced by L. Chua and L. Yang [1a, 1b] has been widely studied due to its interesting features in performing principally image processing tasks. A **CNN** as basically an array of locally interconnected analog processing elements, or cells, operating in parallel, whose dynamic behaviour is determined by the cell connectivity pattern (neighbourhood extent) and a set of configurable parameters. The time evolution of the state of a cell **c** in a N×M-cell CNN is described by the differential equation:

$$\begin{aligned}\tau\frac{dx_c(t)}{dt} &= -x_c(t) + g_c(t)\\ &= -x_c(t) + \sum_n A_{n-c}y_n(t)\\ &\quad + \sum_n B_{n-c}u_n + I\end{aligned}$$

$1 \leq n, c \leq N.M; n \in N_R(c); |x_n(0)| \leq 1; |u_n| \leq 1;$

where **n** denotes a generic cell belonging to the neighbourhood of cell **c**, $\mathbf{N_R(c)}$, with radius equal to R. $N_1(c)$ is the set of 3×3 cells centred in **c** ($N_1(c) = \{c-N-1, c-N, c-N+1, c-1, c, c+1, c+N-1, c+N, c+N+1\}$), $N_2(c)$ the set of 5×5 cells centred in **c**, and so on. $\mathbf{x_c}$ is the state of cell **c**, $\mathbf{y_n}$ is the output from each cell **n**, defined in terms of the nonlinear function:

$$y_n = f(x_n) = \frac{1}{2}(|x_n+1| - |x_n-1|) \qquad y_n \in [-1, 1]$$

$\mathbf{u_n}$ is the input to the cell **n**, **I** is an offset term, and the matrices **A** and **B** are called **feedback** and **control templates** respectively. Depending on the values of the cloning template components, the offset term, and the initial states, the resulting **CNN** is configured or "programmed" to perform a given processing task on the inputs.

Since the publication of the two papers of Chua and Yang, a number of VLSI approaches have been proposed to approximate the CNN model as well as other network models inspired by the original one. Various working chips have also been reported implementing CNNs with either fixed or programmable templates.

On looking at the published designs and experimental results [2–12] (considering primarily analog ones), we can find a wide variety of circuit approaches for CNN cells, depending on the basic circuit techniques used and the way the signals are processed in time (continuous or discrete). Most of the experimental results reported to date refer to the processing of optical information, although in practice, only a few CNN chips integrate on-chip photo-sensors (focal-plane solutions [7, 11, 12]).

The experimental density of programmable CNN chips with embedded photo-sensors ranges between a few and tens of cells per square millimeter, while the chips designed for specific image processing tasks may reach densities of hundreds of cells per square millimeter (if they carry out relatively simple tasks). These densities allow the integration, on a typical 1 cm × 1 cm chip, of CNNs to process images from about a thousand to a few tens of thousands of pixels, depending on the programming features of the networks. If the CNN circuit does not include the photo-sensors the major bottleneck may lie in the way the inputs have to be supplied to the network, which greatly affects the overall speed of the CNN. This probably reduces interest in the CNN implementation itself, since its parallel processing capabilities cannot be fully exploited, and other alternative sequential approaches would probably provide similar performance at lower cost.

Focal-plane CNN chips can provide compact and fact solutions to applications, such as the processing of written characters, where large images and/or high pixel densities are not required. In order to implement large CNNs in VLSI, while maintaining their parallel processing features, alternative approaches capable of coping with the input (and output) problem must be considered. Multiple chip analog implementations would enable, with the present microelectronic technology and without the above integration density constraints, the building of very large CNNs, provided the following two topics are satisfactorily addressed:

- Efficient communication by analog signals (inputs, outputs and inter-chip signals).
- Analog computation independent of process related variations and drifts of circuit characteristics in order to obtain matched behaviour between cells in different chips.

In this paper, three alternative analog implementations of CNNs are described [13]:

- A low-power programmable CNN with embedded photo-sensors.
- A compact fixed-template approach based on single polarity current signals.
- A CNN circuit devised to implement very large CNNs, which uses time multiplexing of digital pulses to represent analog signals and whose dynamics depend only on local ratios of circuit parameters.

The first two approaches, devised for focal-plane chips implementing the basic CNN model, are briefly described, together with experimental results taken from integrated prototypes, in Sections 2 and 3 respectively. The third implementation is more appropriate for very large multiple chip CNNs and is described in Section 4. It allows robust CNN-type networks to be built for spatial and temporal processing of visual information with well controlled time constants at different time scales.

2. Low-Power Programmable CNN Chip with Embedded Photo-Sensors

Figure 1 shows schematically the circuit implementation of a programmable CNN cell. The state equation of cell **c**, belonging to a CNN with M rows and N columns, implemented by the circuit in Fig. 1 [12] is:

$$
\begin{aligned}
C_X \frac{d\,Vx_c(t)}{dt} &= -Iy_c(t) + Ig_c = -Iy_c(t) \\
&\quad + \sum_n S_{An-c} \frac{I_{An-c}}{I_L} Iy_n(t) \\
&\quad + \sum_n S_{Bn-c} \frac{I_{Bn-c}}{I_L} \left(Iu_n \frac{I_L}{I_{LB}} \right) + I \\
Iy_c &= I_L \tanh \frac{k(Vx_c - V_R)}{2U_T}; \; 1 \le c \le MN; \\
& n \in N_R(c); \; Iy_n(t), \; Iu_n \frac{I_L}{I_{LB}} \in [-I_L, +I_L] \qquad (1)
\end{aligned}
$$

All the differential pairs work in weak-inversion, and lateral bipolar transistors are used to distribute the programmable parameters and the current units. The programmable parameters are set by external voltages. For example, the current $\mathbf{IA_{N+1}}$, affected by the sign $\mathbf{SA_{N+1}}$, is used to set the component $\mathbf{A_{N+1}} = \mathbf{SA_{N+1}(IA_{N+1}/I_L)}$ of the feedback template **A**, obtained from the common external voltage $\mathbf{VA_{N+1}}$. The basic difference with respect to the original Chua-Yang model is in the loss term, which depends on the cell output $\mathbf{y_c}$ instead of on the internal cell state $\mathbf{x_c}$. This simplified model is a particularization of the ISR CNN model (Improved Signal Range) formalized in [4] and [11]. The VLSI implementations described in [3, 5, 7] can also be identified as particular cases of such a model. Moreover, a sigmoid-like function, instead of the conventional piecewise-linear function, is used to generate the cell output. The low-current operation of the circuits allows the use of small photo-sensors and reduces power consumption.

The layout of the fabricated 8 × 8 CNN chip based on the circuits in Fig. 1 is shown in Fig. 2. Using a

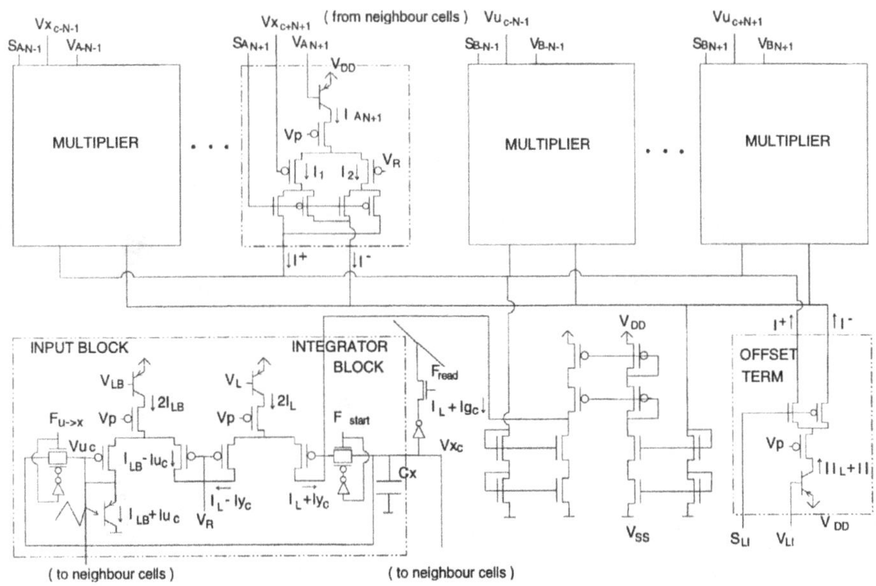

Fig. 1. CMOS implementation of the programmable CNN cell.

1.2μm CMOS process an integration density of 10.7 cells per square millimeter is obtained with 11 programmable parameters (9 feedback parameters, one control parameter and the offset term). The lateral bipolar transistors occupy about 25% of the cell area. Thus, density can be easily improved by implementing the same circuits with a BICMOS process (20% less area is required with the 1.2μm BICMOS process of the same foundry). Only one metal layer has been used for interconnections leaving the second metal layer to protect the circuitry from the incident light. Therefore, an additional metal layer would greatly reduce the cell area due to the large area devoted to interconnection in the present implementation, as can be seen in the layout in figure 2. The chip has been tested in a number of processing tasks with images directly projected onto it, performing either single or multiple sequential programming steps on the same input image: border detection, directional border detection, shadow creation in one, two or three directions, scratching of objects, replacing objects by the smallest rectangles covering them, etc. [16].

Figure 3 shows test results of multiple sequential programming step applications. The network configuration for the different tasks is defined by the parameters presented on the left of the figure. The test results of the implemented 8×8 CNN captured from the chip by a data acquisition board can also be seen on the left. The oscillograms show the cell outputs of a single column and the control signals of the network. $F_{u \rightarrow x}$ enables the CNN input, while the rising transitions of the signal F_{start} activate the network operation for the different processing tasks.

An experimental power consumption of between a few and tens of microwatts per cell has been measured (depending on the network programming) for unit currents ranging from 50nA to 200nA (12μW for a Connected Component Detection, CCD, with a current unit I_L = 150nA, 16μW for a border extraction with I_L = 200nA). The time required by the network to reach its final state is also dependent on the specific processing task. Measured response times are between a few microseconds and a few hundreds of microseconds for the above mentioned unit current range. The worst-case times correspond to tasks in which several cell state transitions take place before the final output is obtained (as for example in the CCD, for which a time of about 150μsec is required using a unit current of 150nA, while a border extraction takes 6μsec with I_L = 200nA).

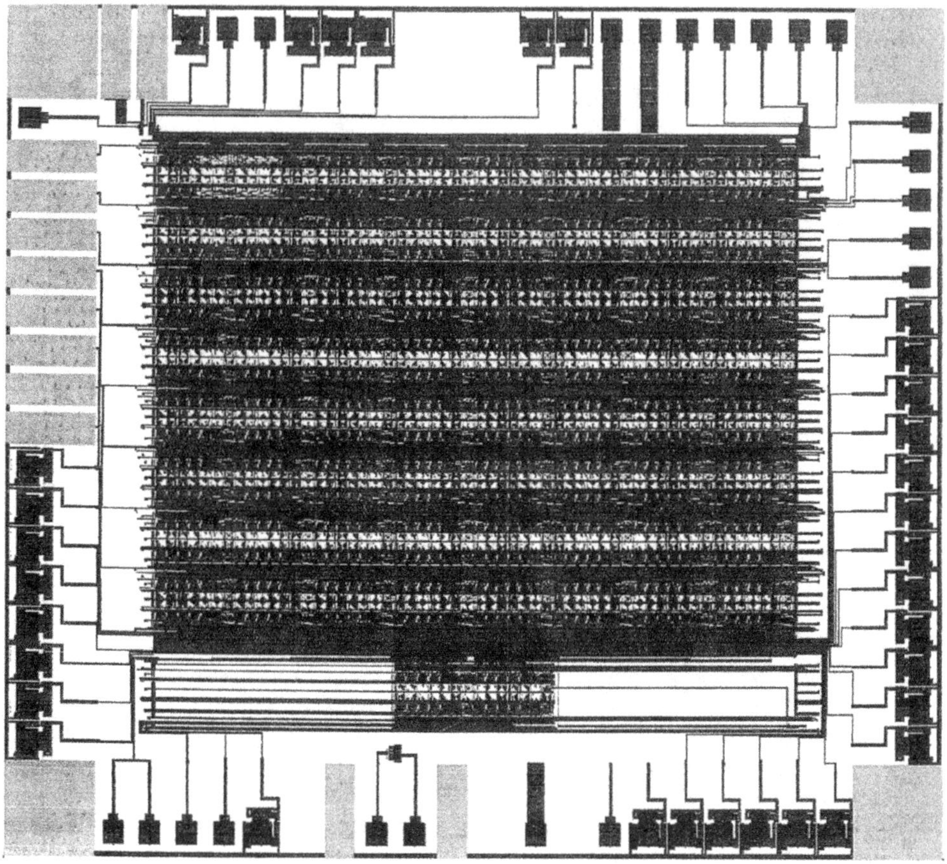

Fig. 2. Layout of the 8 × 8 CNN fabricated chip.

3. Compact CMOS Unipolar Current-Mode CNN Implcmcntation

The current-mode implementations of CNNs usually work with both positive and negative currents. The use of unipolar signals can provide a reduction in silicon area (see also [10] and [15]). In the current mode implementation here described, besides the reduction in the number of transistors required for replication of the unipolar signals, an additional simplification is obtained by the use of simpler circuits to implement the output pseudo-linear function. In order to operate with only positive states a reference shift current I_R is added to the state current of each cell, Ix_c, such that $I_R \geq |I_{RN}|$; with I_{RN} being the negative limit of the state current. According to [1a], $|I_{RN}|$ has a known value for each template configuration:

$$|I_{RN}| = I_L + \left(\sum_n |A_{n-c}| + \sum_n |B_{n-c}|\right) I_L - I \quad (2)$$

Moreover, in order to work with positive inputs and outputs, the current I_L that limits these variables is added; that is: $0 \leq I_L + Iu_c \leq 2I_L$, and $0 \leq I_L + Iy_c \leq 2I_L$.

In terms of the above shifted currents, the state equation of the Chua-Yang CNN model can be written as:

$$\tau \frac{d\,Ix_c}{dt} = -(I_R + Ix_c) + I_R + Ig_c(t)$$

Multiple Sequential Programming steps	*OSCILLOGRAM*
Network Configuration I_L=300nA	

1) Noise $A=2\ B=0\ I=0$

2) Horizontal Scratching [16]

$$A=\begin{bmatrix}0 & 0 & 0\\ 0 & 2a & 0\\ 0 & -2a & 0\end{bmatrix}\quad B=a\quad I=-a\quad a=1$$

3) CCD [14]

$$A=\begin{bmatrix}0 & 1 & 0\\ 0 & 2 & 0\\ 0 & -1 & 0\end{bmatrix}\quad B=0\quad I=0$$

Test results of the 8x8 CNN

2)Scratching

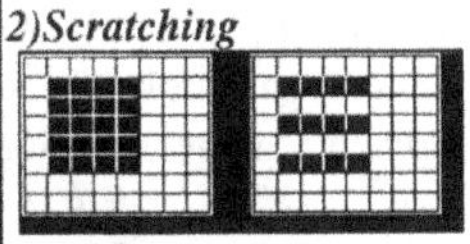

3)CCD

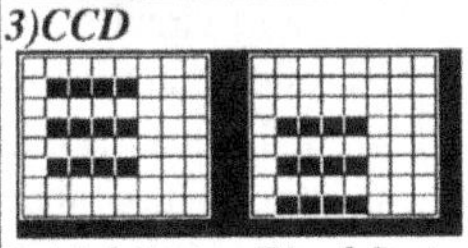

Initial State Final State

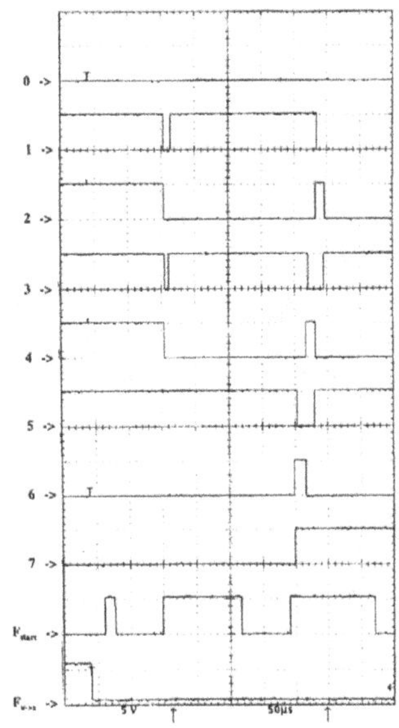

3rd column

Network Configuration I_L=100nA	

1) Noise $A=2\ B=0\ I=0$

2) Replacing objects by the smallest rectangles covering them [16].

$$A=\begin{bmatrix}a & 0 & a\\ 0 & 0 & 0\\ a & 0 & a\end{bmatrix}\quad B=0\quad I=a\quad a=6$$

Test results of the 8x8 CNN

2) Replacing objects by rectangles

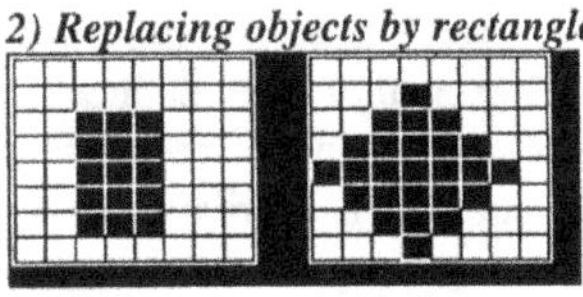

Initial State Final State

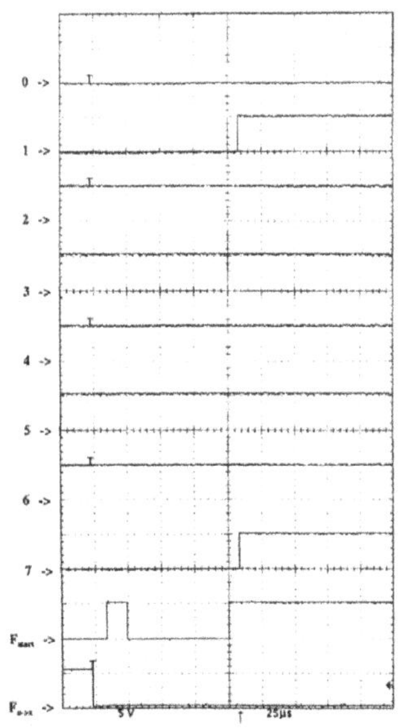

4th column

Fig. 3. Test results of various sequential programming step applications.

$$= -(I_R + Ix_c) + \sum_n (I_L + Iy_n(t))A_{n-c} + \sum_n (I_L + Iu_n)B_{n-c} + I_{off}$$

where

$$I_{off} = I_R + I - \left(\sum_n A_{n-c} + \sum_n B_{n-c}\right) I_L \qquad (3)$$

Figure 4 shows schematically the proposed CMOS implementation of a CNN cell with the above state equation.

A CNN chip for CCD has been designed, fabricated and tested [16], based on the above circuits but with the loss term depending on the output instead of the state of the cell. Figures 5 and 6 show the circuit implementation and layout of a single cell, respectively. Each cell uses only 16 transistors (with cascoded current mirrors as replicators), a density of 230 cells per square millimeter (without photo-sensors) is obtained. If photo-sensors are included, a cell density of 160 cells per square millimeter will be obtained. These density are considerably better than those obtained in the previous CCD-CNN current-mode implementations [2, 7, 11]. Using a power-supply voltage of 3V and a limit current of $2I_L = 1.2\mu$A, a time of $2.3\mu s$ is needed to obtain the CCD computation in a row of 8 cells, and a worst-case power consumption of 13μW per cell has been measured.

4. CMOS Circuits for Multiple Chip CNNs Using Spikes

As in vertebrate visual systems, where the preprocessed visual information coming from the retina is transferred to cortical areas by means of spikes, artificial VLSI vision systems might consist of a dense artificial retina (a dense array of specific photo-sensors performing a basic adaptive contrast of spatial and temporal visual information) whose outputs are transferred to multiple cellular-type layers. These networks, working in continuous time and in parallel, extract relevant spatial and temporal features that must be combined to carry out higher level form and motion processing tasks. To do this, an appropriate interchip communication scheme must be used. It has been demonstrated that non-arbitrated address-driven communications schemes are very efficient in this context [17, 18]. They use the rate of asynchronously produced short pulses (spikes) to represent analog signals, and the time multiplexing of addresses on a fast digital bus to transfer a high number of such signals along a small number of wires. In this way, in contrast with a sequential scanning of images, the relevant visual information is treated preserving its temporal integrity. Typical pulse widths of tens of nanoseconds and interspike times of milliseconds are used obtaining a small collision probability. The circuits here described process directly the signals represented by spike rates implementing the model in Fig. 7. It can be considered as an extension of the basic CNN model in which either linear (current sources representing template components) or shunting terms (conductances) depend on the instantaneous frequency of the respective input spikes. The CMOS circuit implementation of a single positive term or excitatory synapse is shown in Fig. 8 [19]. A dual version of this circuit implements an inhibitory one. Figure 9 is a layout of the integrated synapse including a section of the "membrane capacitance" (C_x in Fig. 7) for a CMOS process of 1.2μm. The same circuit can approximate either a frequency controlled current source (FCI) or a frequency controlled conductance (FCG), depending on the way the vertical dashed-line connection in Fig. 8 is made. For each input pulse, the capacitor C_i^+ is suddenly discharged and then a current is injected on the output node during the time required to charge C_i^+ up to either V_x (shunting terms) or to the reference voltage $V_{r,i}$ (linear terms). For a given input spike frequency F_i^+, the contribution of the synapse to the "membrane potential V_x" can be approximated by [19]:

$$FCG\colon \frac{dV_X}{dt} = \frac{K_i^+ . C_i^+}{C_X} \cdot (V^+ - V_X) \cdot F_i^+;$$

$$FCI\colon \frac{dV_X}{dt} = \frac{K_i^+ . C_i^+}{C_X} \cdot (V^+ - V_{r,i}) \cdot F_i^+ \qquad (4)$$

Assuming that the passive decay conductances $G+$ and $G-$ (see Fig. 7) are also implemented by similar FCG excitatory and inhibitory circuits controlled by a constant frequency reference F_R, the following general equation is approximated when a set of linear and shunting synapses is considered:

$$\frac{dV_x}{dt} = -\frac{G_R}{C_X}(V_X - V_R) + \sum_{j,exc} \frac{I_j^+(F_j^+)}{C_X} - \sum_{j,inh} \frac{I_j^-(F_j^-)}{C_X} + \sum_{i,exc} \frac{g_i^+(F_i^+)}{C_X}(V^+ - V_X)$$

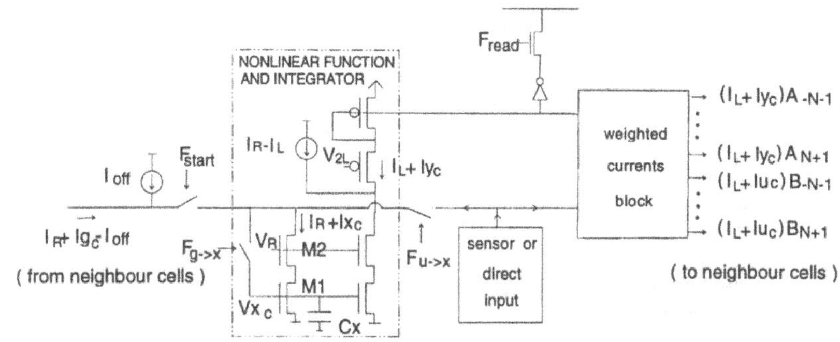

Fig. 4. Circuit implementation of the current-mode fixed-template CNN cell.

- **Parameters [14]:**

$$A = \begin{bmatrix} 0 & 0 & 0 \\ 1 & 2 & -1 \\ 0 & 0 & 0 \end{bmatrix}$$

$B = 0 \quad I = 0 \quad x_b = -1$

u_b = *don't care*

- **Initial state:**

$x_c(0)$=image

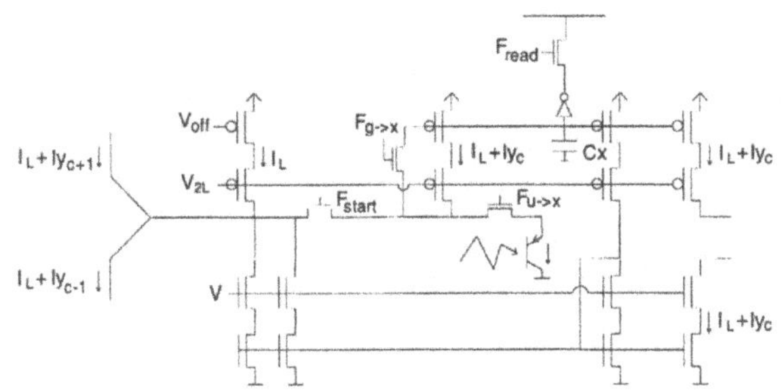

$$\frac{dIy_c}{dt} = -(I_L+Iy_c) + I_L+Ig_c(t) = -(I_L+Iy_c) + \sum_n (I_L+Iy_n(t))\, A_{n-c} + \sum_n (I_L+Iu_n)\, B_{n-c} -$$

$$\text{where} \quad I_{off} = -I_L + \left(\sum_n A_{n-c} + \sum_n B_{n-c}\right) I_L - I$$

Fig. 5. Circuit implementation of a fixed template CNN cell for CCD using unipolar signals.

$$-\sum_{i,inh} \frac{g_i^-(F_i^-)}{C_X}(V_X - V^-) \qquad (5)$$

where:

$$G_R = G^+(F_R) + G^-(F_R);$$

$$V_R = \frac{G^+.V^+ + G^-.V^-}{G^+ + G^-};$$

$$g_i^{+/-}(F_i^{+/-}) = K_i^{+/-} \cdot C_i^{+/-} \cdot F_i^{+/-};$$
$$I_j^+(F_j^+) = K_j^+ \cdot C_j^+ \cdot (V^+ - V_{r,j});$$
$$I_j^-(F_j^-) = K_j^- \cdot C_j^- \cdot (V_{r,j} - V^-)$$

Thus, the circuits implement an extension of the CNN model by the inclusion of the shunting terms. The first three terms in eqn. (5) model the conventional decay and template terms in the CNN model. As can be seen, all temporal contributions to the state variable V_X are defined by external references and local ratios of currents (K terms) and capacitances, which are more controllable in VLSI than absolute values. The output module in Fig. 7 is implemented by a circuit [20] that performs a pseudo-linear conversion of V_X to a spike rate, whose output frequency is also defined by external

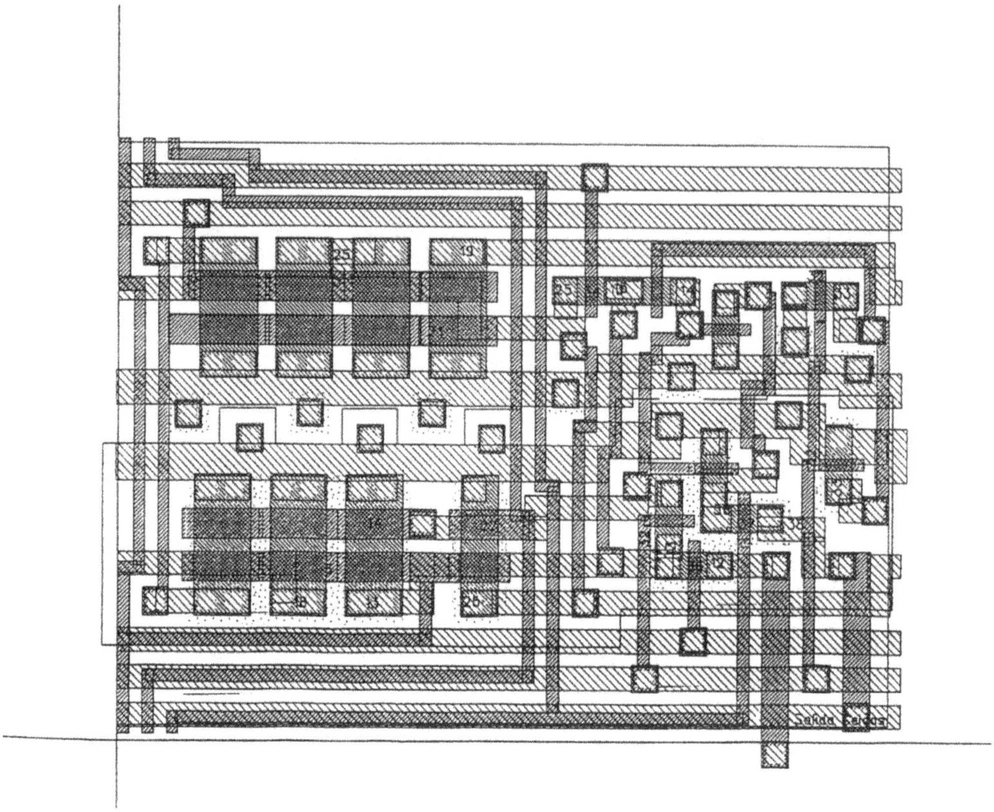

Fig. 6. Layout of one cells of the integrated current-mode CNN for CCD (80 × 54μm^2) with the 1.2μm CAE process of AMS.

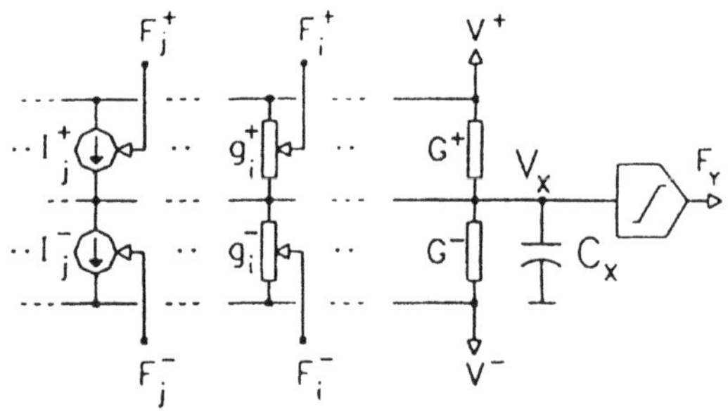

Fig. 7. Extended CNN cell model in which inputs and output are defined by spike rates.

references and local ratios:

$$F_Y(V_X) = \frac{V_X - V_{\min}}{V_{DD}} \cdot \frac{C_0}{C_F} \cdot K \cdot F_0 \qquad (6)$$

$V_{\min}$ and F_0 are external voltage and frequency references; C_0 and C_F are internal capacitances of the spike generator, and K is a local current ratio.

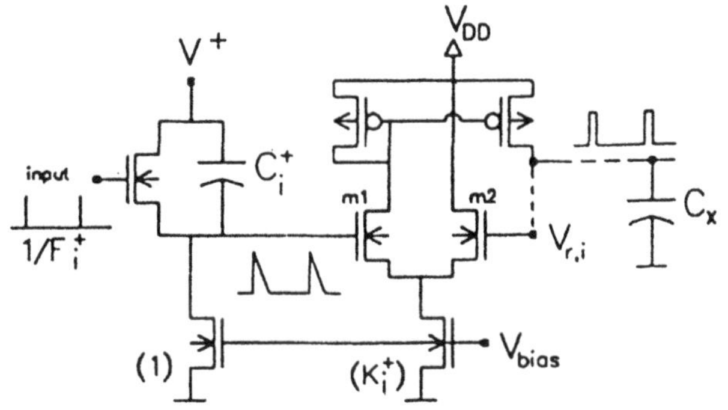

Fig. 8. CMOS circuit implementation of an excitatory synapse in the model of Fig. 7.

Figure 10 shows experimental measurements illustrating the good linear behaviour, the control of the time constant, and the variability of the template components. In this experiment V^+ is set to 3.5V and the three traces V_{X1}, V_{X2}, and V_{X3}, taken from different chips. They correspond, respectively, to reference voltages $V_{r,i} = 2$V, 2.5V and 3V (in eqn. 4) representing template values of 3, 2 and 1, when one linear synapse is excited with a constant input frequency. Additional experimental results illustrating features of the model resulting from the shunting terms can be found in [20]. In the present implementation, the circuits operate in the range of micro to milliseconds; thus, instead of being considered fast "accelerators" of traditional CNNs, they should be considered good approximators of the continuous dynamics of an extended computing model, by means of which many large processing layers of cellular networks can be built, with full parallel operation, while maintaining a good control of the time constants at different time scales.

Figure 11 shows a multiple chip configuration where a dense artificial retina would perform parallel sensing and a spatio-temporal contrast. A non-arbitrated address-driven communication scheme is used to interconnect the different modules through fast digital buses. These interchip connections can be controlled by a digital optional block where the weights of programmable synapses can be modified by modulating the instantaneous frequencies of pulses that would arrive at the corresponding destination cells. In this case, the figure presents the possibility of having several cellular-type layers processing, in parallel, different features of the retinal output. Fig. 12 shows how a single large CNN layer can be split into several chips where local encoder/decoders are used to connect neighbour cells in different ICs while global encoder/decoders achieve the transmission of input/output signals. Both multiple chip architectures can be combined to perform complex analog processing tasks, maintaining parallelism and asynchrony in the whole network.

5. Concluding Remarks

The paper summarizes three different VLSI approaches to cellular structures.

First an analog CMOS implementation of CNN with the following features: (a) Embedded photosensors for direct capturing of input images, thus exploiting the potential advantage of CNNs for parallel image processing. (b) Low-current operation, which allows the use of small photo-sensors and reduces power consumption. (c) The parameters defining the processing task to be carried out by the CNN are modifiable by means of external signals, which enable different processing tasks to be performed on the same or different input images.

Second, a fixed-template CMOS implementation based on the use of single-polarity current-mode signals has been shown, which is useful to carry out specific processing tasks on images with a higher pixel density and speed than allowed by the programmable approach. A considerable improvement in area is obtained due to the reduction in the number of transistors required to signal replication, and the simplification of circuits to generate the pseudo-linear output function. Based on these circuits a chip with 8 cells for CCD has been integrated and tested with a density of 230 cells per square millimeter (excluding the photo-sensors).

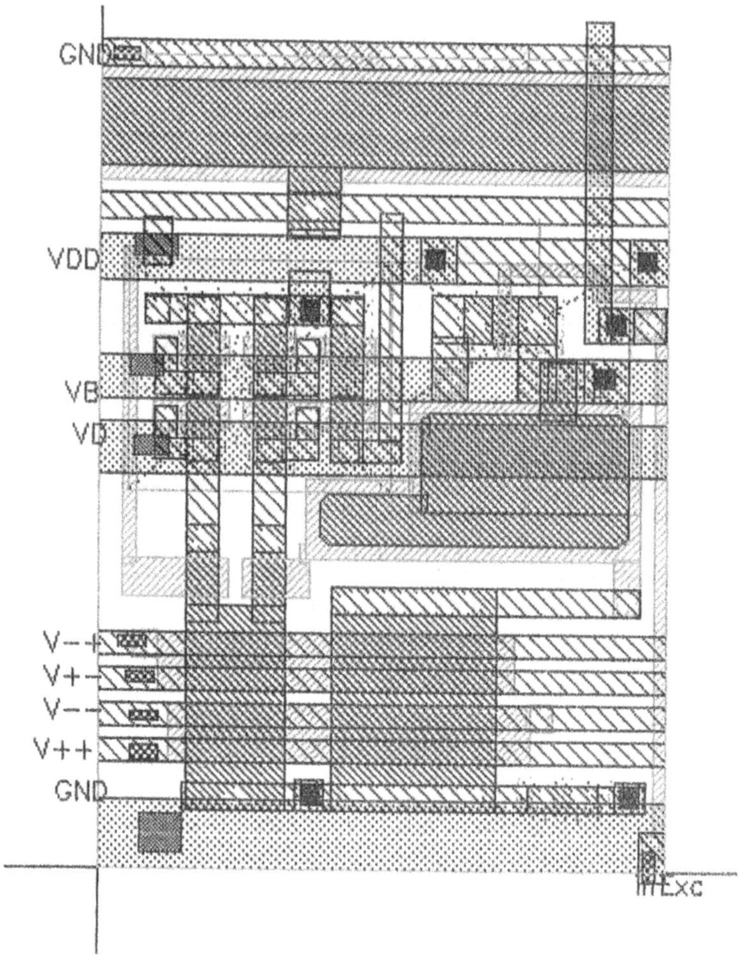

Fig. 9. Layout of the excitatory synapse of Fig. 8. This occupies 60 × 60 μm^2 excluding the section of membrane capacitance.

Third, a circuit approach to approximate the continuous dynamics of an extended CNN model, in which input and output analog signals are represented by asynchronously produced short pulses, is proposed as a viable alternative for the implementation of large analog cellular structures. Due to the goal control of the circuit parameters, which are defined in terms of local ratios and to the use of an appropriate digital communication scheme that greatly reduces the number of required interconnections, with these of circuits a large cellular array of analog cells can be split into multiple chips, preserving the dynamics of the network as a whole.

The third VLSI approach corresponds to CNN circuits devised to implement very large CNNs with an accurate control of the transient and the steady-state neural responses, both being defined by process-independent local ratios. The design of the circuits with well controlled time constants provides the possibility of using cellular-type layers combined with artificial retinas to perform temporal processing tasks like novelty detection and motion processing. Interneuron communication is achieved by means of very short pulses multiplexed in fast digital buses, preserving features of the information. Adjustable weights for

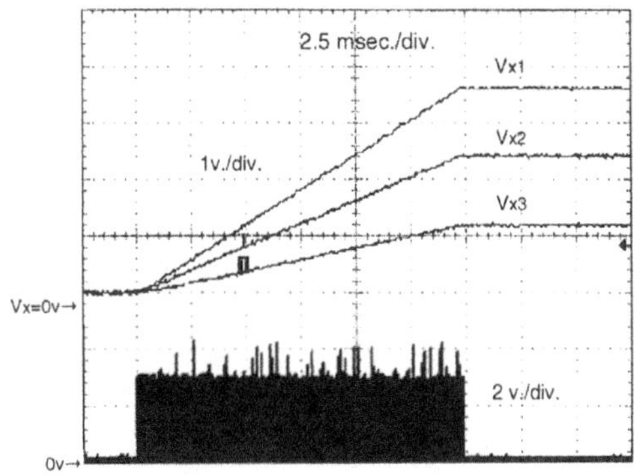

Fig. 10. Temporal evolution of the states of integrated cells, with a constant input applied to a linear synapse, for three template values.

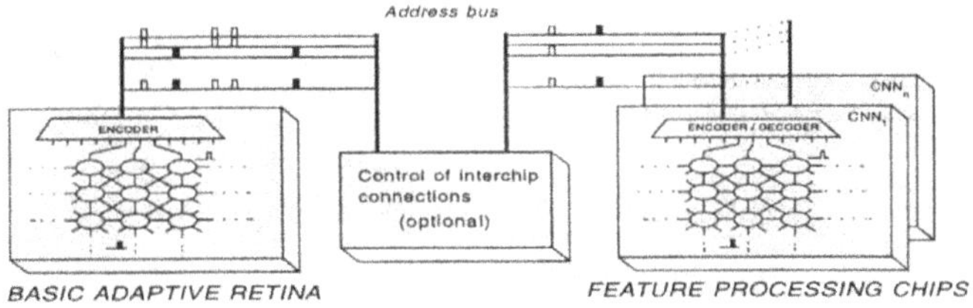

Fig. 11. Multiple chip CNN implementation: system architecture of a low level vision system based on the scheme in [21].

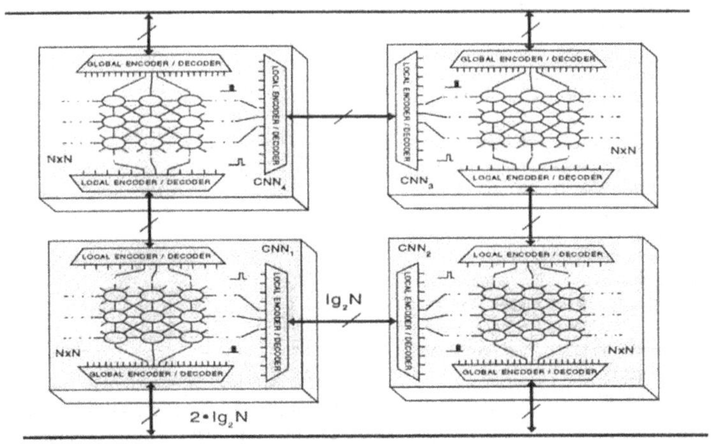

Fig. 12. Multiple chip CNN implementation: CNN split into several chips using additional local address encoders/decoders to interconnect neighbour cells in different chips.

the shunting terms (FCG) can be implemented either using variable ratio current mirrors (parameters $K_i^{+/-}$ and $K_j^{+/-}$ in eqns. 4 and 5) or modifying, by means of digital circuitry, the inter-neuron spike frequencies. The weights of the linear terms (FCI) can be modulated through the adjustable voltage references $V_{r,i}$ in eqns. 4 and 5.

Acknowledgements

This work has been partially supported by the EC Project ERBCHBGCT920027, and the Spanish CICYT Project TIC96-0634.

References

1a. L. O. Chua and L. Yang, "Cellular neural networks: Theory." *IEEE Transactions on Circuits and Systems* 35 (10), pp. 1257–1272, Oct. 1988.
1b. L. O. Chua and L. Yang, "Cellular neural networks: Applications." *IEEE Transactions on Circuits and Systems* 35 (10), pp. 1273–1290, Oct. 1988.
2. J. M. Cruz and L. O. Chua, "A CNN chip for connected component detection." *IEEE Transactions on Circuits and Systems* 38 (7), pp. 812–816, July 1991.
3. H. Harrer, J. A. Nossek, and R. Stelzl, "An analog implementation of discrete-time cellular neural networks." *IEEE Transactions on Circuits and Systems* 39 (3), pp. 466–476, May 1992.
4. A. Rodríguez Vázquez, S. Espejo, R. Domínguez Castro, J. L. Huertas, and E. Sánchez Sinencio, "Current-mode techniques for the implementation of continuous and discrete time cellular neural networks." *IEEE Transactions on Circuits and Systems Part I* 40 (3), pp. 132–146, March 1993.
5. J. E. Varrientos, E. Sánchez-Sinencio, and J. Ramírez-Angulo, "A current-mode cellular neural networks implementation." *IEEE Transactions on Circuits and Systems Part I* 40 (3), pp. 147–156, March 1993.
6. M. Anguita, F. J. Pelayo, A. Prieto, and J. Ortega, "Analog CMOS implementation of a discrete time CNN with programmable cloning templates." *IEEE Transactions on Circuits and Systems Part I* 40 (3), pp. 215–219, March 1993.
7. S. Espejo, A. Rodríguez-Vázquez, R. Domínguez-Castro, J. L. Huertas, and E. Sánchez-Sinencio, "Smart-pixel cellular neural networks in analog current-mode CMOS technology." *IEEE Journal of Solid-State Circuits* 29 (8), August 1994.
8. F. Sargeni and V. Bonaiuto, "High performance digitally programmable CNN chip with discrete templates," in *Proceedings of the CNNA-94*, Rome, 1994, pp. 67–72.
9. A. Paasio, A. Dawidziuk, K. Halonen, and V. Porra, "Current mode cellular neural network with digitally adjustable template coefficients," in *Proceedings Microneuro '94*, IEEE Comp. Soc. Press., Turin, 1994, pp. 268–272.
10. P. Kinget and M. S. J. Steyaert, "A programmable analog cellular neural network CMOS chip for high speed image processing." *IEEE Journal of Solid-State Circuits* 30 (3), March 1995.
11. S. Espejo, "VLSI design and modeling of CNNs," PhD. Dissertation, University of Sevilla, March, 1994.
12. M. Anguita, F. J. Pelayo, F. J. Fernandez, and A. Prieto, "A low-power CMOS implementation of programmable CNNs with embedded photo-sensors." To appear in *IEEE Transactions on Circuits and Systems Part I.*
13. M. Anguita, F. J. Pelayo, E. Ros, D. Palomar, and A. Prieto, "VLSI implementations of CNNs for image processing and vision tasks: Single and multiple chip approaches," invited talk at the *4th IEEE International Workshop on Cellular Neural Networks and their Applications*, June, 1996.
14. T. Matsumoto, L. O. Chua, and H. Suzuki, "CNN cloning template: Connected component detector." *IEEE Transactions on Circuits and Systems* 37, pp. 633–635, May 1990.
15. A. J. Schuler, M. Brabec, D. Shubel, and J. A. Nossek, "Hardware-oriented learning for cellular neural networks," in *Proceedings of CNNA-94*, Rome, Italy, December 1994, pp. 183–188.
16. M. Anguita, "Implementación de arquitecturas VLSI para redes neuronales celulares (CNNs)," Ph.D. Dissertation, Departamento de Electrónica y Tecnología de Computadores, Universidad de Granada, June 1996.
17. A. Mortara and E. A. Vittoz, "A Communication Architecture Tailored for analog VLSI Artificial Neural Networks: Intrinsic Performance and Limitations." *IEEE Transactions on Neural Networks* 5 (5), May 1994.
18. A. Mortara, A. E. Vittoz, and Ph. Venier, "A communication scheme for analog VLSI perceptive systems." *IEEE Journal of Solid State Circuits* 30 (6), pp. 660–669, June 1995.
19. F. J. Pelayo, E. Ros, P. Martin-Smith, F. J. Fernández, and A. Prieto, "A VLSI approach to the implementation of additive and shunting neural networks," in *Lecture Notes in Computer Science*, Springer-Verlag, July 1995, vol. 930, pp. 728–735.
20. F. J. Pelayo, E. Ros, X. Arreguit, and A. Prieto, "A bio-inspired VLSI neural model using spikes." Sent to *Analog Integrated Circuits and Signal Processing*, Kluwer Academic Publishers.
21. X. Arreguit and E. A. Vittoz, "Perception systems implementation in analog VLSI for real-time applications," in *Proceedings of the PerAc'94*, Lausanne, Switzerland, 1994, pp. 170–180.

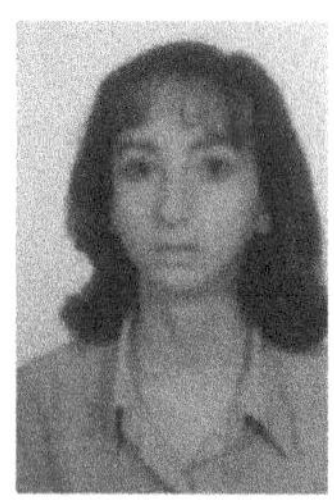

Mancia Anguita received a B.Sc. degree and a Ph.D. in Computer Science from the University of Granada, Spain, in June 1991 and July 1996. Since 1991 she is a teaching assistant at the Department of Electronics and Computer Architecture at that University. Her research interests lay in parallel computer architecture and analog VLSI circuit implementations.

Francisco J. Pelayo, born in Granada, Spain, in 1960, received the B.Sc. degree in electronic physics in 1982, the M.Sc. degree in electronics in 1983, and the Ph.D. degree in 1989, all from the University of Granada, Spain.

In September 1983, he joined the Departamento de Electrónica y Tecnología de Computadores at the University of Granada, where he is currently an Associate Professor. He has worked in the areas of multiple-valued logic, VLSI design and test, artificial neural networks and fuzzy systems. His current research interest lies in the fields of VLSI artificial neural networks, and bio-inspired processing systems.

Eduardo Ros received the B.Sc. Degree in Electronic Physics in 1993, and Electronic Engineering in 1996, both from the University of Granada, Spain. He joined the Departamento de Electrónica y Tecnología de Computadores 1993, where he is currently Research Assistant. His research interests include bio-inspired VLSI neural systems.

David Palomar was born in Almuñécar (Spain) on 1972. He was graduated on Computer Sciences by the University of Granada in 1995. Now he enjoys a fellowship from Spanish E.M. and is working toward his Ph.D. degree at the Department of Electronics and Computer Technology of the University of Granada. He is interested in analog VLSI implementation of neural systems, with special attention to vision related systems.

Alberto Prieto received a B.Sc. degree in Electronic Physics in 1968 from the Complutense University (Madrid) and a Ph.D. degree from the University of Granada, Spain, in 1976. From 1969 to 1970 he was at the Centro de Investigaciones Técnicas de Guipuzcoa and at the E.T.S.I Industriales of San Sebastián. From 1971 to 1984 he was Director of the Computer Centre, and from 1985 to 1990 Dean of the Computer Science and Technology studies of the Univ. of Granada. He is currently a Full Professor in the Dept. de Electronica y Tecnologia de Computadores.

His research interests are in the areas of artificial neural networks, multiple-valued circuits, microprocessor-based systems, and VLSI design and testing. Dr. Prieto has been nominated member of the IFIP WG 10.6 (Neural Computer Systems). He is a member of INNS, AEIA, AFCET, IEEE and ACM, and Chairman of the Spanish Regional Interest Group of the IEEE Neural Network Council.

Analog Integrated Circuits and Signal Processing, 15, 277–290 (1998)

Architecture and Design of 1-D Enhanced Cellular Neural Network Processors for Signal Detection

MICHELLE Y. WANG,[1] BING J. SHEU,[1] THEODORE W. BERGER[2], WAYNE C. YOUNG[1] AND AUSTIN K. CHO[1]

[1] *Department of Electrical Engineering, University of Southern California, Los Angeles, CA 90089-0271;* [2] *Department of Biomedical Engineering, University of Southern California, Los Angeles, CA 90089-1451*

Abstract. One-dimensional cellular array processor architecture and design for neural-based partial response (PR) signal detection are presented. Analog parallel computing approaches have many attractive advantages in achieving low power, low cost, and faster processing speed by its uniquely coupled parallel and distributed processing nature. In this paper, we describe the maximum likelihood sequence estimation (MLSE) algorithm for PR signals, the enhanced Cellular Neural Network (CNN) processor array architecture to realize the detection algorithm, and system performance evaluation. Analytical models and simulations on a design example of the detector have been employed to demonstrate the advantages of this scalable VLSI architecture. A processing rate of 265 Mbps was achieved for a prototype detector on a silicon area of 5.14 mm by 5.81 mm is a 1.2μm CMOS technology. The processing rate can be beyond 1Gbps if it is implemented in the same amount of silicon area by using 0.5μm CMOS technology. Such promising results clearly demonstrate the ability to meet the needs in future high speed data communication by VLSI realization of maximum likelihood sequence detectors based on the enhanced cellular neural network paradigm.

1. Introduction

Partial response signaling representation is a class of transmission coding schemes with minimized degradation for many high speed communication systems. For example, the quaternary partial-response class-IV (QPRIV) scheme was used for fiber distributed data interface (FDDI) [1] in twisted pair communication cables. Many industrial implementations of digital Viterbi detector at the data rate around 150 Mbps or higher were recently reported for the maximum likelihood sequence detection (MLSD) of the recording channels in magnetic disk drivers [2, 3, 4]. Analog circuit approaches [5] have been considered to provide better alternative solutions to many barriers in digital implementation. In digital Viterbi detection implementation, there are many summation, comparison, switching, and storage operations. These operations have to be executed either serially at a low data rate or in parallel with significant increase in hardware resource. In addition, a high-resolution A/D converter is required. The A/D converter usually occupies quite a large silicon area.

Cellular neural networks are a class of analog networks with locally connected identical computing cells, called neurons. It was introduced by Chua and Roska [6] for a 2-dimensional mesh-grid analog array processors for image processing. There are several working neuronal models for cellular neural networks. Chua-Yang's model [7] is suitable for image processing and was used in universal machine [6]. The original Chua model has out-of-bound equilibrium points which could be difficult to implement in very large scale integration (VLSI) chip. However, it can be improved by a Full Signal Range (FSR) model proposed by Rodriguez-Vasquez [8].

The proposed *enhanced cellular neural network* is an analog neuromorphic VLSI processing system. A state-constrained neuronal (SCN) model [9] which effectively solves the out-of-bound equilibrium problem and the inaccurate state resistance problem, is used to realize parallel-distributed optimization. The scalable architecture has the following important features:

(1) pure neuromorphic sensory information processing, with no conventional A/D converter,

(2) pipelined operation to exploit data parallelism, and

(3) allowing high-speed sampling. It includes the organization for a current-mode VLSI neuron, the timing information for the analog processor array, and a set of mapping rules to realize the neural computation on the target VLSI hardware.

In this paper, the architecture for PRML detection algorithms is presented. The corresponding maximum likelihood sequence estimation (MLSE) algorithm is first described. In section 3, the mapping method from the MLSE algorithm to processing array is reported. In section 4, detector design example is given, which can be extended to different partial response signaling schemes. The timing diagram and block diagram are also described.

2. Maximum Likelihood Sequence Detection Algorithm

Partial response signaling representation can produce a target response at the detector for data communication system given a rectangular binary symbol at the input [10]. Partial response signaling deliberately generates inter-symbol interference (ISI) and a degree of correlation between transmitted symbols or signal levels. Thus, it can be referred to as correlative coding. It is used in baseband systems and various carrier formats including frequency modulation and spread-spectrum broadcasting.

In addition to data communication systems, it is popular to apply partial response maximum-likelihood (PRML) detection scheme for the data receiver design of the tape or disk reading channel in magnetic recording systems.

The block diagram of a communication system is shown in figure 1. The waveform of the coded data stream is determined by the characteristics of the coder. The coder characteristic functions for different classes of partial response signaling are listed in Table 1 [11]. A bipolar digital data sequence $d(k)$ is fed into the coder. The partial response signaling coder, which could be a delay-impulse response filter or a magnetic reading channel, produces a code stream $td(k)$ for transmission as $td(k) = \sum_i d(i)f(k-i)$, where $f(k)$ is the kth component in the impulse response and its z-transform is $F(Z)$. Note that $f(k) = 0$ if $k < 0$.

The transmitted data stream is sent to an analog channel which contains unwanted noise source, and interferences. In this study, an additive white Gaussian noise is assumed. The received data sequence $r(k)$ after channel equalization takes the form of

$$r(k) = \sum_i d(i)f(k-i) + n(k), \qquad (1)$$

where $n(k)$ is a random white Gaussian noise sequence.

The channel equalizer and the clock recovery circuit blocks are to process analog values. In fact, analog channel equalizer and clock recovery circuit were suggested to reduce cost and electrical power consumption [5].

The optimal detection problem is to estimate the data sequence based on the observed sequence $\boldsymbol{r_n}$. The maximum likelihood sequence estimator selects an estimation vector $\boldsymbol{u_n} = \{u(0), u(1), u(2), \ldots, u(n-1)\}$, whose elements are in the range of $[-1, 1]$, as a best estimate, that maximizes the conditional *a posterior* probability $p(\boldsymbol{r_n} \mid \boldsymbol{d_n})$, where $\boldsymbol{r_n} = \{r(0), r(1), r(2), \cdot, r(n-1)\}$.

If the noise is i.i.d. Gaussian with the same standard deviation for all symbols in a data sequence, this optimization problem can be solved by minimizing the quadratic likelihood metric function $M(\boldsymbol{u_n})$,

$$M(\boldsymbol{u_n}) = \sum_{j=1}^{n} \left(r(j) - \sum_i u(i)f(j-i) \right)^2. \qquad (2)$$

Because an exponential function is a monotonically decreasing function, the instance which has the minimum exponent has the minimum value.

3. Algorithm Mapping to Enhanced Cellular Neural Network

The neuronal model used for enhanced cellular neural networks is the state-constrained Neuronal (SCN) model. It is a mathematical dynamic system model which can be implemented in voltage-mode or current-mode design avoiding the out-of-bound equilibrium point problem. It is suitable for optimization applications because of its high linearity, accuracy, and guaranteed stability at the grid points in the solution space of a symmetric network. The diagram of an SCN neuron is shown in figure 2. A state-constrained neuron cell C_i, where i is the index number of the cell, includes a neuronal body to perform integration on the state variable

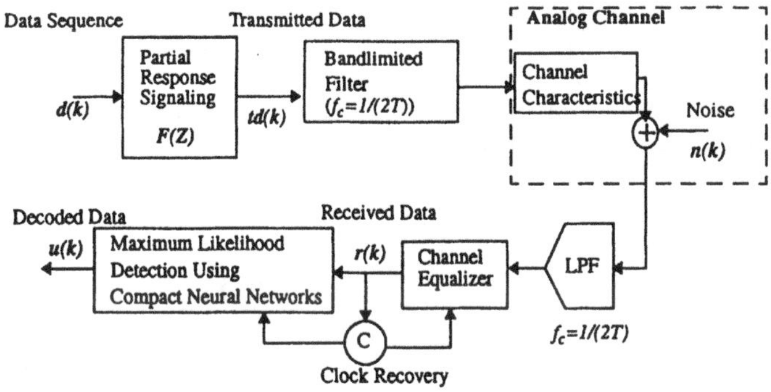

Fig. 1. A communication system with cellular neural network detector.

Table 1. System characteristic functions for partial response (PR) signaling systems [12].

PR coder characteristic function $F(Z)$	PR Classification	Output levels
$1+Z^{-1}$	PR1 (duobinary)	3
$1+2Z^{-1}+Z^{-2}$	PR2	5
$2+Z^{-1}-Z^{-2}$	PR3	5
$1-Z^{-2}$	PR4 (modified duobinary)	3
$1+Z^{-1}-Z^{-2}-Z^{-3}$	EPR4 (extended PR4)	5
$1-2Z^{-2}+Z^{-4}$	PR5	5

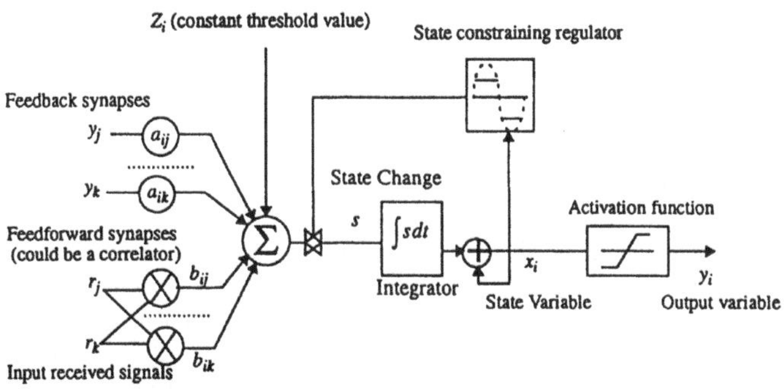

Fig. 2. State Constrained Neuronal (SCN) model for neuron cell C_i.

x_i; an activation function to produce output y_i that is routed to neighboring nodes; several synaptic connections a_{ij} to neighbors, and external input coefficients b_{ik}, where j and k are the index numbers of the neighboring cells. The dynamics of the state variable x_i is governed by

$$\tau\frac{dx_i}{dt} = \begin{cases} \sum_{i\in N(C_i)} a_{ij}y_i + \sum_{k\in I(C_i)} b_{ik} + Z_i & -1 \le x_i \le 1 \\ 0 & x_i \le -1 \text{ or } x_i \ge 1 \end{cases} \quad (3)$$

where $y_i = f(x_i)$ is output of the activation function, $N(C_i)$ is the set of neighboring cells which has synaptic connections with C_i, and $I(C_i)$ is the set of input nodes which provides data to determine external input coefficients b_{ik}'s into the state variable x_i. In this paper, a non-linear piece-wise function

$$f(x) = \frac{1}{2}(|x+1| - |x-1|), \tag{4}$$

is considered as the activation function.

Now consider a network without the threshold component, i.e. $Z = 0$. A Lyapunov energy function could be found for the network if the solution trajectory stays within the solution grids enforced by the SCN model. The energy function for the network is

$$E = -\frac{1}{2}\mathbf{X}^{\mathbf{T}}\mathbf{AX} - \mathbf{BX}. \tag{5}$$

This energy function has a gradient of

$$\begin{aligned} \nabla_x E &= -\mathbf{AX} - \mathbf{B} = -(\mathbf{AX} + \mathbf{B}) \\ &= -(\mathbf{Af(X)} + \mathbf{B}) = -\tau \frac{\mathrm{d}\mathbf{X}}{\mathrm{dt}} \end{aligned} \tag{6}$$

$$-1 \leq x_i \leq 2, \text{ for every } i.$$

where $f(\mathbf{X}) = \mathbf{X}$ if all components of $\mathbf{X}$ are within $[-1, 1]$ and $\mathbf{A} = \mathbf{A_T}$ is a symmetric matrix.

Thus, the energy function satisfies the Lyapunov condition because

$$\frac{\mathrm{dE}}{\mathrm{dt}} = \nabla_x \mathrm{E}\frac{\mathrm{d}\mathbf{X}}{\mathrm{dt}} = -\tau\left(\frac{\mathrm{d}\mathbf{X}}{\mathrm{dt}}\right)^2 \leq 0. \tag{7}$$

Due to the inequality relation, the trajectory of the state vector of a cellular neural network is a path with monotonically decreasing energy function. Minimization problems can be mapped onto a cellular neural network for the optimal solution with the use of this energy function. Detailed analysis on the optimization theory for cellular neural network can be found in [12]. In principle, an energy function can be found for all symmetric synaptic networks because all the eigenvalues for the connection matrix are real-valued. In this design, the cost function E can be defined as

$$\begin{aligned} E &= \frac{1}{2}M(\mathbf{u_n}) - \sum_{k=0}^{n-1} r(k)^2 \\ &= \frac{1}{2}\sum_{k=0}^{n-1}\sum_{j=0}^{n-1}\sum_{m=0}^{n-1} u(j)u(m)f(k-j)f(k-m) \\ &\quad - \sum_{k=0}^{n-1}\sum_{i=0}^{n-1} r(k)f(k-i)u(i). \end{aligned} \tag{8}$$

In the first term, a matrix $\mathbf{D}$ can be defined by its elements so that

$$\begin{aligned} &\frac{1}{2}\sum_{k=0}^{n-1}\sum_{j=0}^{n-1}\sum_{m=0}^{n-1} u(j)u(m)f(k-j)f(k-m) \\ &\quad = \frac{1}{2}\mathbf{U}^{\mathbf{T}}\mathbf{DU}. \end{aligned} \tag{9}$$

and the element in jth row and mth column of $\mathbf{D}$ is

$$d_{jm} = \sum_{k=0}^{n-1} f(k-j+1)f(k-m+1). \tag{10}$$

Notice that $d_{mj} = d_{jm}$, and matrix $\mathbf{D}$ is symmetric. This is true for the partial response signaling schemes. For the second term in the energy function E, the coefficient vector can be defined as

$$b(i) = \sum_{k=0}^{n-1} r(k)f(k-i). \tag{11}$$

To map the maximum likelihood detection algorithm onto the enhanced cellular neural network, it is required to define a connection matrix $\mathbf{A}$ for the network. The connection matrix $\mathbf{A}$ is defined as $-\mathbf{D}$. The dynamic equation for the network is

$$\tau\frac{d\mathbf{X}}{dt} = \mathbf{AX} + \mathbf{B}. \tag{12}$$

and the energy function becomes

$$E = -\frac{1}{2}\mathbf{X}^{\mathbf{T}}\mathbf{AX} - \mathbf{BX}. \tag{13}$$

The state variable vector $\mathbf{X}$ has a direct mapping to the estimator vector $\mathbf{U}$ of the data sequence.

The connection matrix can be linearly scaled for larger networks. All the connection matrices are symmetric. A partial response maximum likelihood sequence detecting problem can be solved with direct mapping using the connection matrix and correlators.

The quadratic likelihood metric function of an enhanced cellular neural network is similar to the likelihood metric function used in the Viterbi detectors [5, 13]. The network stability is influenced by the input pattern and the noise level. In order to find a suitable operation time for the neural network, computer simulation was conducted. A 4-neuron network for the maximum likelihood sequence detection of the class IV

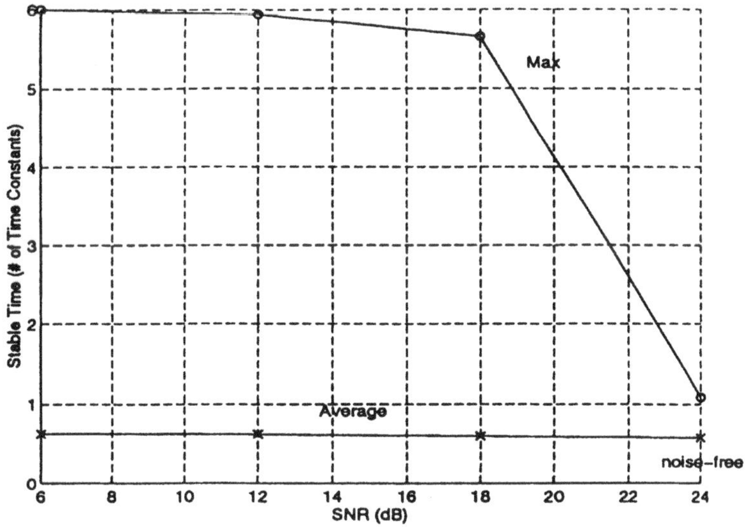

Fig. 3. The effect of noise level on the network stable time in term of time constants.

partial response signaling scheme was simulated with Matlab tools on a SPARC-20 workstation. For noise-free data input streams, 16 input patterns are studied. The simulation result indicates that the network can reach the stable state within 2 time constants. In this 4-neuron network, 1,000 random data patterns for each of the 16 input combinations were fed into the network. The simulation results are plotted in figure 3. The detection operations for all of the possible input combinations of signal-to-noise ratio ranging from 6dB, 12dB, to 18dB can be stabilized within 6 time constants. Notice that the SNR for a magnetic reading channel of a hard disk drive ranges from 12dB to 18dB [5]. The worst cases found in the simulation under different noise levels and different inter-symbol interference level L are all bounded within 6 time constants. This implies an excellent property of this network that the processing time is well constrained over different intersymbol interference levels. This is a clear advantage over digital serial Viterbi approach which has a time complexity in order of $o(M2^L)$, where M is size of the symbol set. If wireless communication carrier is used, M is 2 for BPSK cases, while M is 4 for QPSK cases.

The chip area complexity, in term of the number of neurons, for an enhanced cellular neural network is determined by the constraint length of the detectors which is called interference level L. The required network for PRML detection should has $L + 1$ neurons. Therefore, the area complexity of the proposed architecture is on the linear order of $o(L)$.

To compare the proposed solution with a digital Viterbi maximum likelihood detector, simulation of the bit-error-rate (BER) under various signal-to-noise ratios (SNRs) was performed. The results are shown in figure 4. In this experiment, 1,000 random data sequence runs were used as the input to a 20-neuron processor array. The simulation results indicate that the difference of the maximum likelihood detection result between the 20-neuron network and the digital Viterbi detector is within 5–15% for different SNR values.

4. System Architecture

In this section, the scalable VLSI architecture for PRML detection is described. A 64-neuron design example is shown in figure 5. It uses the G_m-C style neuron, as shown in figure 6, which includes a state capacitor, a transconductive multiplier to enable the annealing capability (or an operational transconductive amplifier for a non-annealed neuron), multiple synapses, and a few multipliers for the correlator. In this design example, the communication detector contains two time-multiplexed analog-processor modules, each for one

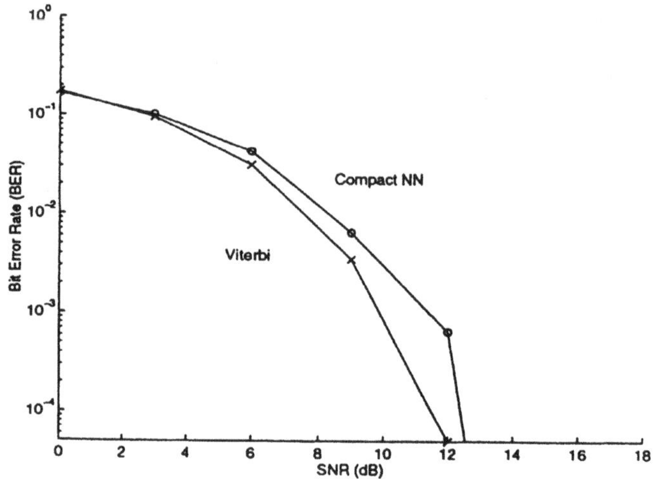

Fig. 4. Bit error rate performance for a 2-neuron processor array.

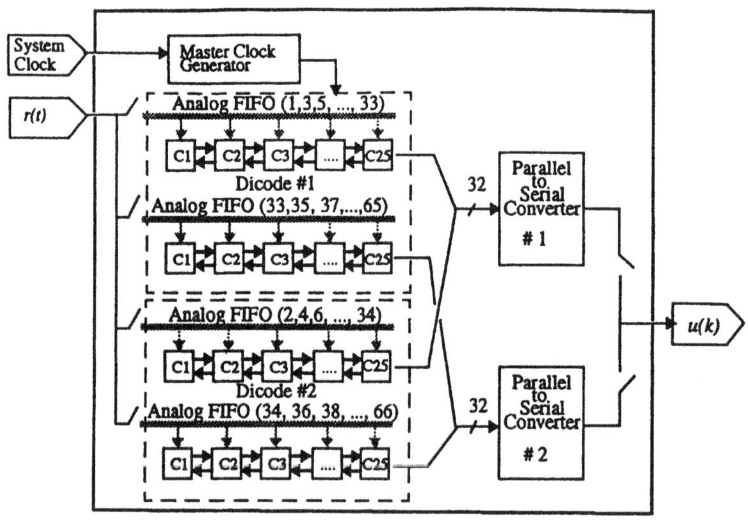

Fig. 5. Block diagram of a cellular neural network for PRIV MLSD (a 64-neuron example).

dicode. Each array processor module contains two processing arrays which operate in the pipelined format. It is designed for the PRML detection in the class IV partial response scheme which can be used for the recording channel in hard disk drives. Each of the prototyping processor arrays contains 16 neurons. There are two parallel-to-serial converters to convert the parallelized data sequence back to the serial format.

A detector contains the following architectural components:

(1) a master clock generator,

(2) four VLSI analog processor arrays,

(3) four analog FIFOs to buffer the received signal $r(t)$, and

(4) two parallel-to-serial converters. The master clock generator is actually a controller for the operation

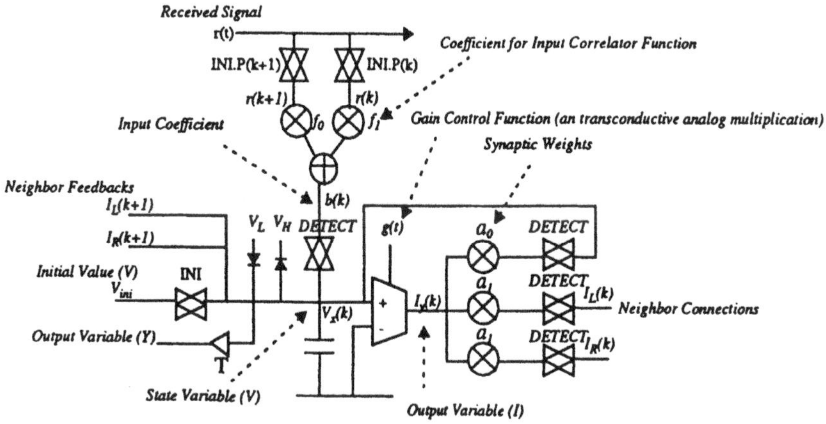

Fig. 6. A current-mode neuron for cellular neural networks.

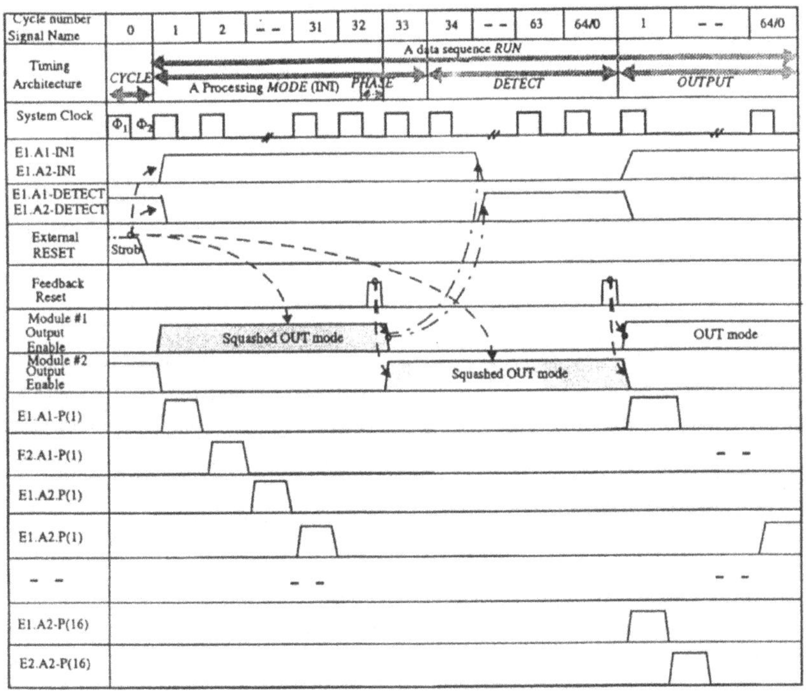

Fig. 7. Timing diagram for an enhanced cellular neural network with 64 neurons.

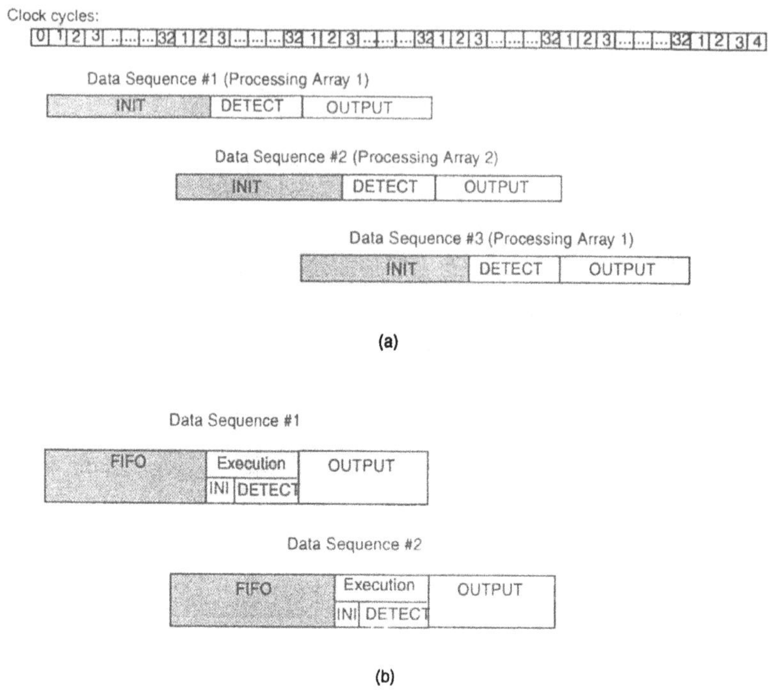

Fig. 8. Pipelined operation steps for the enhanced cellular neural network VLSI cores. (a) Two-array design for a processing module. (b) Single-array for a processing module.

of the whole detector. It generates the synchronization signal to the processing arrays to select between the initialization mode and the detection mode, and the control signal for the parallel-to-serial converter. The analog array processors are organized according to the enhanced cellular neural network paradigm. The parallel-to-serial converter can be used to convert the output sequence from a parallel format into a serial format to reduce the pin-count in practical VLSI implementation and make the detector compactible with surrounding serial digital signal processing hardware.

The timing diagram for the whole processing detector is shown in figure 7. The system clock is extracted from the data stream by a clock recovery circuit and used to generate the two-phase clocks by the master clock generator. The reset strobe resets the master clock generator to an initial value and, thus, resets the system. The master clock generator produces an INI signal in a length of 34 clock cycles for the processor array 1 in both the dicode 1 and dicode 2. In each of the 66 cycles, a sampling phase clock is generated to sample each data point in $r(k)$ sequence. 66 data samples are required for a data sequence run of the two dicodes in the PRIV signaling system with a level of interference of 1 ($L = 1$), since the number of the required data samples for a data sequence run is $F_{out} + L$. Thus, the length of the detection mode (DETECT) has only 30 cycles. The network results are in binary format and are buffered by the parallel-to-serial converter. In the output (OUT) mode, the data are shifted out. In a data sequence run, the input data samples are distributed into the processing arrays according to the number of elements in the analog FIFOs as shown in figure 6. Similarly, the 64 output bits in the parallel-to-serial converter are re-assembled according to that order. For a non-PR IV detector, only one dicode, with two processing array, is needed. If each array has F_{out} neurons, the $F_{out} + L$ clock cycles are needed for the INI mode, and $F_{out} - L$ cycles for the DETECT mode.

Since the operation can be partitioned into the INI mode and the DETECT mode, a two-stage time-

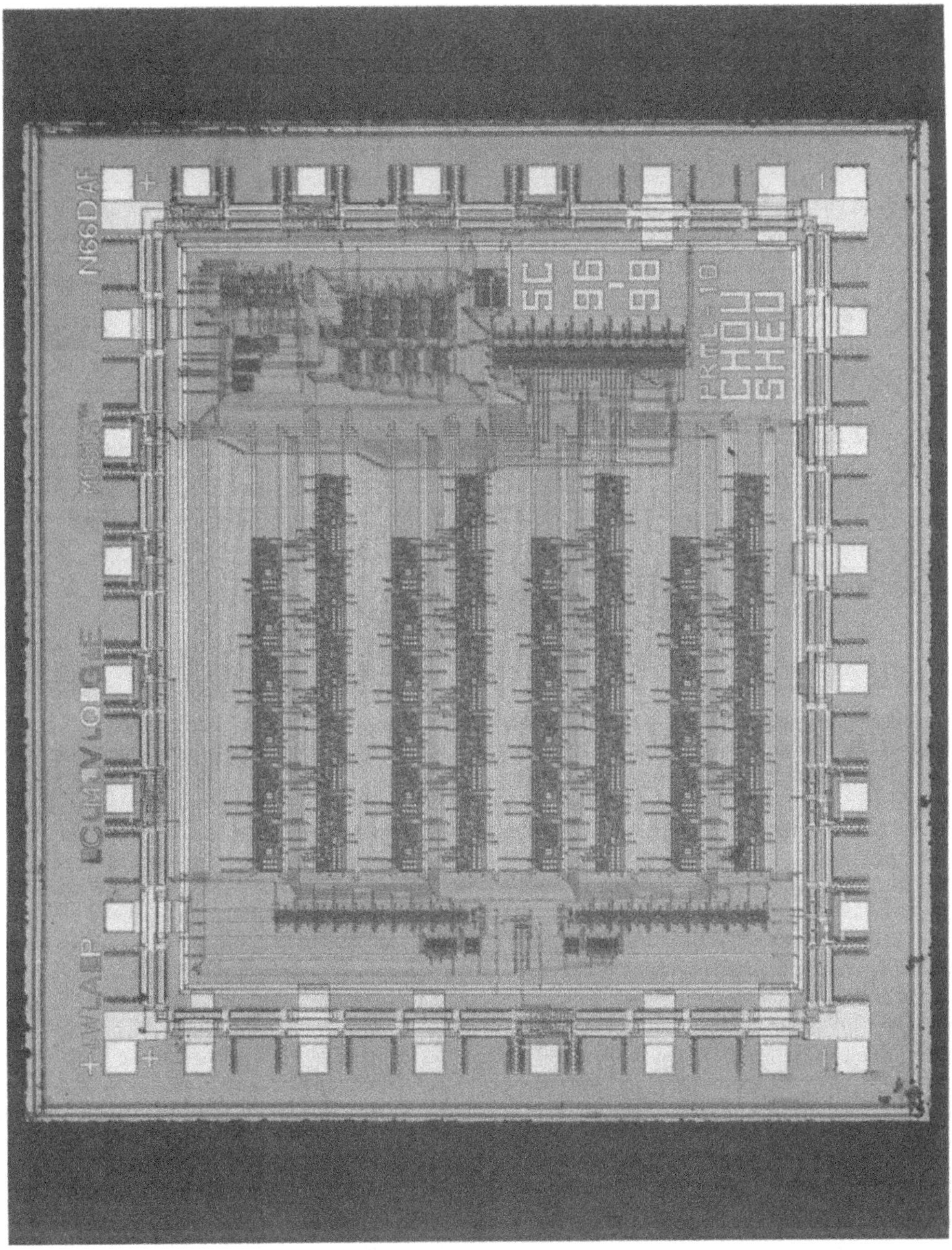

Fig. 9. Layout of the 20-neuron PRIV detector.

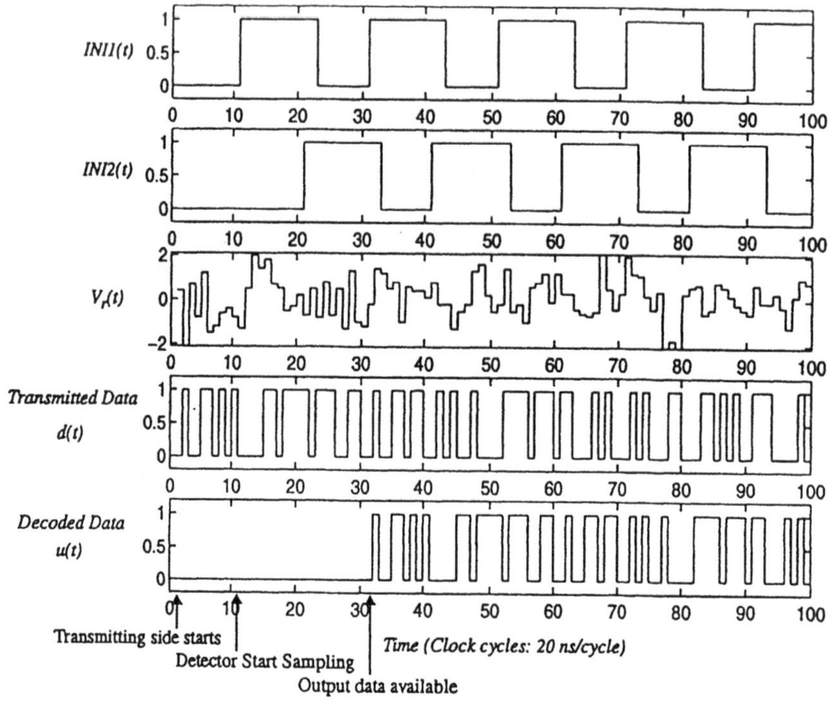

Fig. 10. Behavior of the 20-neuron PRIV detector.

multiplexing pipelined processing design is suitable for each dicode. The processing steps are shown in figure 8(a). It is also possible to share only one processing array with two analog FIFOs to buffer two incoming signal sequences. The processing steps of the single-array of pipelined operation is shown in figure 8(b). When one data sequence is in operation, the other data sequence including the initial values is accumulated in the second FIFO. However, for the single-array pipelining to work properly, it requires accurate analog-RAM cells and analog data forwarding devices. Notice that there is one bit shared by two consecutive data sequences. Some early versions of the analog-RAM are available [14]. An accurate analog data forwarding device could be a voltage follower which requires the chip area of at least the size of a voltage-mode operational amplifier that could be larger than the current-mode neuron itself. This single-array version can be very efficient when efficient analog RAM technology becomes more mature.

Then to map this detection algorithm onto a VLSI array-processing module, the following design decision and adjustments have to be made:

(1) choice of the processor transconductance (G_m) value;

(2) scaling according to the signal operational range based on the chosen IC technology;

(3) adding proper constraint energy function;

(4) adjustment of the synapse weight values to the range according to selected power supply value; and

(5) tuning the time constant by sizing the state capacitor.

These includes the consideration for the limitations, design rules of the supporting IC technology and the design optimization at the algorithmic level.

A 20-neuron analog processor array prototype for PRIV signal detection was designed in 1.2μm CMOS technology on a silicon area of 5.14mm by 5.81mm. The chip layout is shown in figure 9. The circuit behavior is shown in figure 10. It was simulated in HSPICE [15]. Figure 11 shows the output variables of the processing module with an input data sequence of [00111 01111 01011 00011].

The processing rate of this prototype was 265 Mbps.

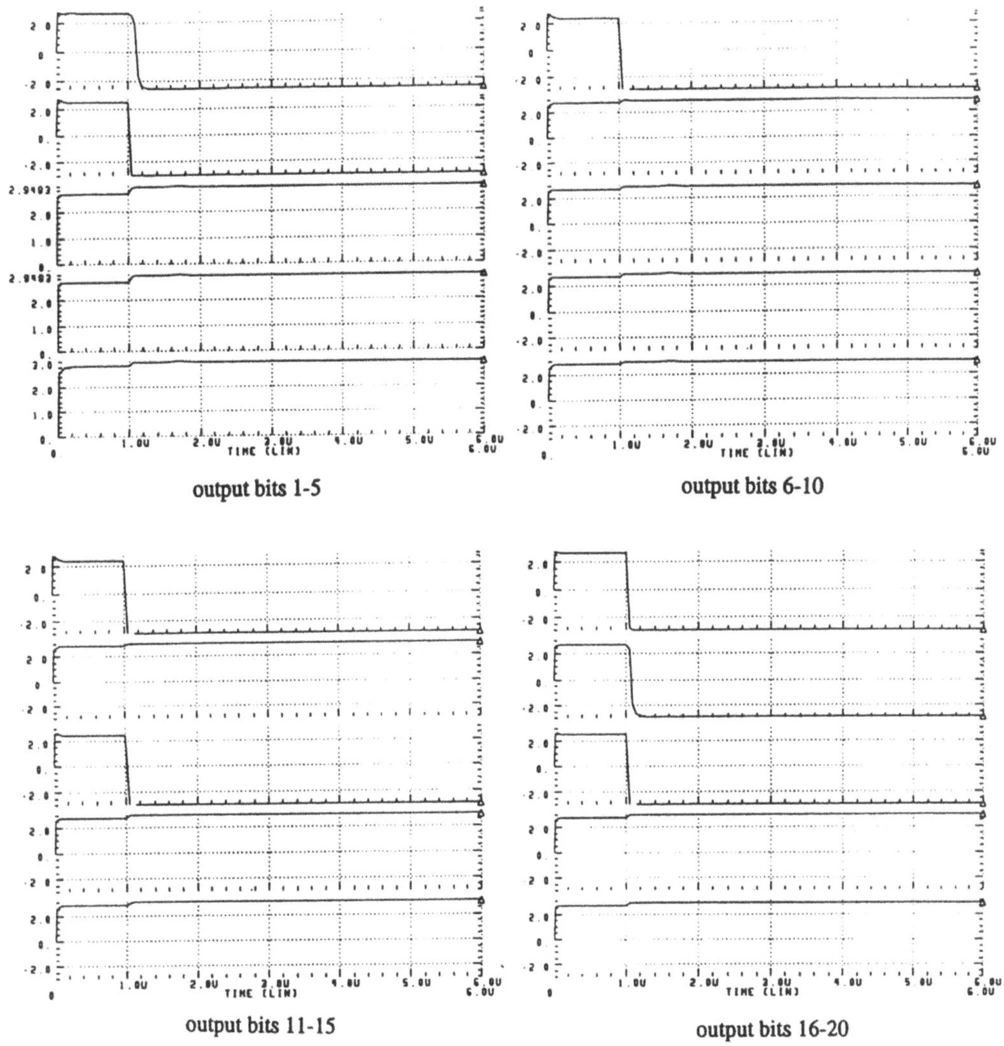

Fig. 11. Hspice simulation results of the 20-neuron PRIV detector.

By adjusting the transconductance (G_m) value and the size of the state capacitors, the processing rate of the same 20-neuron structure can go as high as 1 Gbps. By adding the number of neurons, this processor has a potential processing rate of more than 10 Gbps. Thus, this kind of cellular neural network processor can play a very important role in very high speed communication systems, especially in the optical connected systems, because of the analog nature of the transmitting signals. And the compact structure of this processor is suitable for narrow-structure realization.

5. Conclusion

The proposed enhanced cellular neural network algorithm and architecture are suitable for maximum likelihood detection of partial response signaling method which is widely used in communication systems including magnetic reading channels. Simulation results illustrate that the operation time for a detection operation grows slowly with network size, and the network size grows linearly with the symbol interference level, instead of exponentially like the digital Viterbi algorithm. The architecture has been applied to a prototype chip design for a data processing rate of 265 Mbps with 5.14 mm by 5.81 mm silicon area in a 1.2μm CMOS technology. The proposed VLSI architecture has a potential of 10 Gbps data processing capability with the state-of-the-art IC fabrication technology. This scalable VLSI architecture can be applied to various communication applications which incorporates a partial response signaling scheme.

Acknowledgements

This research was partially supported by research grant from Office of Naval Research under Contract N00014-96-1-0331.

References

1. P. Kabal and S. Pasupathy, "Partial response signaling." *IEEE Trans. on Communications* 23 (9), pp. 921–934, Sept. 1975.
2. S. Mita and Y. Ouchi, "A 150Mb/s PRML chip for magnetic disk drives," in *IEEE Inter. Solid-State Circuits Conference*, San Francisco, CA, Feb. 1996, pp. 62–63.
3. G. T. Tuttle and G. D. Visshakhadatta, "A 130 Mb/s PRML read/write channel with digital-servo detection," in *IEEE Inter. Solid-State Circuits Conference*, San Francisco, CA, Feb. 1996, pp. 64–65.
4. K. Parsi and N. Rao, "A 200 Mb/s PRML read/write channel IC," in *IEEE Inter. Solid-State Circuits Conference*, San Francisco, CA, Feb. 1996, pp. 66–67.
5. R. R. Spencer, "Simulated performance of analog Viterbi detectors." *IEEE Journal on Selected Areas in Communications* 10 (1), pp. 277–299, Jan. 1992.
6. T. Roska and L. O. Chua, "The CNN universal machine: an analogic array machine." *IEEE Trans. on Circuits and Systems—II* 40, pp. 163–173, Mar. 1993.
7. L. O. Chua and L. Yang, "Cellular neural networks: theory and applications." *IEEE Trans. on Circuits and Systems* 35 (10), pp. 1257–1290, Oct. 1988.
8. A. Rodriguez, S. Espejo, R. Dominguez-Castro, J. Huertas, and E. Sanchez-Sinencio, "Current-mode techniques for implementation of continuous- and discrete-time cellular neural networks." *IEEE Trans. on Circuits and Systems—II* 40 (3), pp. 132–146, March 1993.
9. E. Y. Chou, B. J. Sheu, and R. H. Tsai, "A state constrained model for cellular non-linear network optimization," accepted by *IEEE Trans. on Circuits and Systems—I* for publication.
10. H. Lin and D. Messerschmitt, "Parallel Viterbi decoding methods for uncontrollable and controllable sources." *IEEE Trans. on Communications* 41, pp. 62–69, Jan. 1993.
11. G. Wade, *Signal Coding and Processing*. Cambridge University Press: New York, 1994.
12. S. H. Bang, B. J. Sheu, and E. Y. Chou, "A parallel hardware annealing method for optimal solutions of cellular neural networks." *IEEE Trans. on Circuits and Systems—II* 43 (6), pp. 409–421, June 1996.
13. G. D. Forney, Jr., "The Viterbi algorithm." *Proc. IEEE* 61 (3), pp. 268–275, Mar. 1973.
14. E. Franchi, M. Tartagni, R. Guerrieri, and G. Baccarani, "Random access analog memory for early vision." *IEEE Journal of Solid-State Circuits* 27 (7), pp. 1105–1109, July 1992.
15. HSPICE User's Manual, Meta-software, Inc.: Campbell, CA, 1996.

Michelle Yibing Wang received the B.S.E.E. and M.S.E.E. degrees from Tsinghua University, Beijing, China, in 1992 and 1994 respectively. Currently, she is a Ph.D. student in the intelligent machines and multimedia field at University of Southern California. She works in the VLSI Multimedia Laboratory of Electrical Engineering Department. Her current research interests are wireless communication systems, low-power VLSI system design and intelligent microsystems.

Bing J. Sheu was born in Taiwan in 1955. He received the B.S.E.E. degree (Honors) in 1978 from the National Taiwan University, the M.S. and Ph.D. degrees in electrical engineering from the University of California, Berkeley, in 1983 and 1985, respectively.

In National Taiwan University, he was the recipient of the Distinguished Book-Coupon Award for 7 times. In 1981, he was involved in custom VLSI design for a speech recognition system at Threshold Technology Inc., Cupertino, CA. From 1981 to 1982, he was a Teaching Assistant in the EECS Department, UC Berkeley. From 1982 to 1985, he was a Research Assistant in the Electronics Research Laboratory, UC Berkeley, working on digital and analog VLSI circuits for signal processing. In 1985, he joined the faculty in Electrical Engineering Department at University of Southern California and is currently a Professor with a joint appointment in Biomedical Eng. Department. He has been an active researcher in several research organizations at USC including Signal and Image Processing Institute (SIPI), Center for Neural Engineering (CNE), Institute for Robotics and Intelligent Systems (IRIS), and NSF/ERC Integrated Media Systems Center (IMSC), and serves as Chair of Colloquium Series and Publications in IMSC. He serves as the Director of VLSI and Signal Processing Laboratory. Since 1983, he has served as a consultant to the microelectronic and information processing industry. His research interests include VLSI chips and systems, massively paralleled neural networks and image processing, and high-speed interconnects and computing. He is an Honorary Consultant in National Chiao Tung University, Hsin-Chu, Taiwan.

Dr. Sheu was a recipient of the 1987 NSF Engineering Initiation Award and, at UC Berkeley, the Tse-Wei Liu Memorial Fellowship and the Stanley M. Tasheira Scholarship Award. He was also a recipient of the Best Presenter Award at IEEE International Conf. on Computer Design in both 1990 and 1991. He is a recipient of the Best Paper Award of IEEE Transactions on VLSI Systems in 1995, Best Poster Paper Award of World Congress on Neural Networks from International Neural Network Society in 1995, and NASA Certificate of Recognition in 1995. He has published more than 180 papers in international scientific and technical journals and conferences and is a coauthor of the book *Hardware Annealing in Analog VLSI Neurocomputing* in 1991, and the book *Neural Information Processing and VLSI* in 1995 from Kluwer Press and co-editor of *Microsystems Technology for Multimedia Applications* in 1995 from IEEE Press. He served on the Technical Program Committees of IEEE ISCAS, ICASSP, IJCNN, ICCD, CICC. He served as a Guest Editor of IEEE Jour. of Solid-State Circuits for March 1992 and 1993 Special Issues; a Guest Editor on computer technologies for IEEE Transactions on VLSI Systems for June 1993 Special Issue; an Associate Editor of IEEE Transactions on Neural Networks, of Trans. on Circuits and Systems Part I (Express Letters) and Part II (VLSI DSP). At present, he serves as Associate Editor of IEEE Transactions on VLSI Systems; CAS Editor of IEEE Circuits and Devices Magazine, Guest Editor of Multimedia Special Issue on IEEE T-CSVT. He also serves on the editorial boards of J. of Analog ICs & Signal Processing, Kluwer Press; and Int'l Neurocomputing Journal, Elsevier Press. He serves as the Tutorials Chair of 1995 IEEE Int'l Symposium on Circuits and Systems; Technical Program Chair of 1996 IEEE Int'l Conf. on Neural Networks, and Tech. Prog. Chair of 1997 IEEE Int'l Conference on Computer Design; Tech. Prog. Chair of 1997 Int'l Conference on Next Decades of High Technologies, and is currently Editor-in-Chief of IEEE Trans. on VLSI Systems. He was among the key contributors of the widely used BSIM model in the SPICE circuit simulator. He is a member of International Neural Networks Society, Eta Kappa Nu, and Phi Tau Phi Honorary Scholastic Society, and IEEE Fellow.

Dr. Theodore W. Berger received the B.S. (1972) in psychology from Union College, Schenectady, NY and Ph.D. (1976) in physiological psychology from Harvard University, Cambridge, MA. In Union College, he received honors from Phi Beta Kappa, Sigma Xi, departmental honors in psychology, and the George H. Catlin Prize awarded for highest scholastic record and promise for graduate study. In Harvard University, he was awarded an NIMH predoctoral fellowship. During 1976–1977, Dr. Berger worked as an postdoctoral research associate with Dr. Richard F. Thompson, University of California, Irvine. He was a faculty member in psychobiology program, department of psychology and psychiatry, University of Pittsburgh, 1979–1986. He became an associate professor in department of behavioral neuroscience and psychiatry, University of Pittsburgh, 1986–1987 and was promoted to full professor in 1987. In 1992, he became Professor in departments of Biomedical Engineering and Biology (Neurobiology), and Program in Neural, Informational, and Behavioral Science, in University of Southern California.

Dr. Berger received the James McKeen Cattell Award (New York Academy of Sciences), 1978; Alfred P. Sloan Foundation Fellow in Basic Research (Neurosciences), 1978; McKnight Foundation Scholar Award in the Neurosciences, 1980; Research Scientist Development Award (NIMH, Type II), 1981, 1982; Fellow of American Association for the Advancement of Science, 1986, Research Scientist Award (NIMH), 1993. His research involves the complementary use of experimental and theoretical approaches to develop biologically constrained mathematical models of mammalian neural systems. He is Director of Center for Neural Engineering at University of Southern California.

Wayne C. Young was born in Taiwan in 1970. He received the B.Sc. degree in Electrical Engineering from National Taiwan University, Taipei, Taiwan in 1992, and the M.S. degree in Electrical Engineering from University of Southern California in 1995. Currently, he is pursuing the Ph.D. degree in the same field at University of Southern California.

In 1994, he joined the VLSI Multimedia Laboratory at USC and worked as a Research Assistant. During Fall 1995 semester, he was invited to participate in a DARPA-sponsored low-power computing project at USC/Information Science Institute, Marina del Ray, CA. In Spring 1996, he served as a Teaching Assistant for graduate-level course on high performance microelectronic computing & signal processing. In Summer and Fall 1996, he worked as a Senior Engineer at the Physical Optics Corporation, Torrance, CA on industrial-level SBIR innovative project. He is a co-author of 4 journal/conference papers, and serves as Special Assistant in the Colloquium Series of Integrated Media Systems Center, which is an NSF/Engineering Research Center. His current research interests at VLSI Multimedia Laboratory include multimedia technology, emphasizing microelectronic circuits/systems and software implementation, as well as real-world applications for creative business opportunities.

Austin Kwang-Bo Cho received the B.S. (Summa Cum Laude) and M.S. in electrical engineering, from the Sogang University and the Korea Advanced Institute of Science and Technology (KAIST), Korea, in 1992 and 1994, respectively. Currently he is a Ph.D. candidate in Electrical Engineering at the University of Southern California.

From 1994 to 1995, he was a research engineer at the LG Corporate Institute of Technology, Korea. Since 1996, he has been a part-time researcher at Photobit LLC, La Crescenta, CA, as a VLSI design engineer. He has published more than 20 articles on topics in image processing and intelligent information processing. His current research interests are active CMOS image sensors, low-power VLSI system design, and intelligent microsystems.

Analog Integrated Circuits and Signal Processing, 15, 291–314 (1998)

Analog VLSI Circuits for Competitive Learning Networks

H. C. CARD, D. K. McNEILL AND C. R. SCHNEIDER

Department of Electrical and Computer Engineering, University of Manitoba, Winnipeg, Manitoba, Canada R3T 5V6

Received July 15, 1996; Accepted October 24, 1996

Abstract. An investigation is made concerning implementations of competitive learning algorithms in analog VLSI circuits and systems. Analog and low power digital circuits for competitive learning are currently important for their applications in computationally-efficient speech and image compression by vector quantization, as required for example in portable multi-media terminals. A summary of competitive learning models is presented to indicate the type of VLSI computations required, and the effects of weight quantization are discussed. Analog circuit representations of computational primitives for learning and evaluation of distortion metrics are discussed. The present state of VLSI implementations of hard and soft competitive learning algorithms are described, as well as those for topological feature maps. Tolerance of learning algorithms to observed analog circuit properties is reported. New results are also presented from simulations of frequency-sensitive and soft competitive learning concerning sensitivity of these algorithms to precision in VLSI learning computations. Applications of these learning algorithms to unsupervised feature extraction and to vector quantization of speech and images are also described.

Key Words: Artificial Neural Networks, Vector Quantization, Competitive Learning, Analog VLSI Circuits

1. Introduction

There is a current wealth of research activity in the telecommunications and computer industries concerning the role of VLSI technology in ISDN applications and potential products such as hand-held multimedia terminals [1,2]. These portable computers should be capable of two-way multimedia communication with larger computer nodes within a building in the presence of other similar devices, and therefore must contend with a limited available bandwidth, even though the larger nodes may be connected over high-bandwidth optical fiber channels (Fig. 1). Of primary importance in the achievement of these products are efficient data compression algorithms for speech and video signals [3]. It is also required that these tasks be completed with a low battery power budget, and in real time so that the signals do not have to be stored during compression and decompression operations. This paper considers the opportunities for achieving these goals with microelectronic circuits implementing artificial neural network (ANN) models.

Competitive learning (CL) employing VLSI circuits is well suited to the compression of data such as speech and images. Competitive learning is one means of vector quantization (VQ) as the competing units learn to represent the various VQ codebook entries as their respective weight vectors [4,5]. This type of data compression is lossy, but can result in an excellent approximation to the original data provided the codebook size is sufficiently large. As a result, it is appropriate to the compression of speech and images, in which perfect reconstruction of the original signals following decompression is usually not required.

Competitive learning networks are employed both in parallel codebook search, which is a very rapid operation in parallel VLSI implementations, and in the initial learning of the codebooks based upon training data. The latter procedure is time consuming and would normally be performed off-line using workstations rather than on portable or otherwise power-limited computing devices. An intermediate scheme is to perform most of the learning off-line, but to allow the codebook entries (weight vectors) to continue to slowly adapt or fine-tune themselves to changes in their environment. For example a streetcorner voice-activated banking machine that was trained on a corpus of speech in the absence of noise may have sufficient computational resources to adapt to nonstationarities in its individual environment such as traffic noise. Similar methods may be employed in fine-tuning hearing instruments to individual hearing problems or in adapting video compression to particular noise environments. Self-organizing

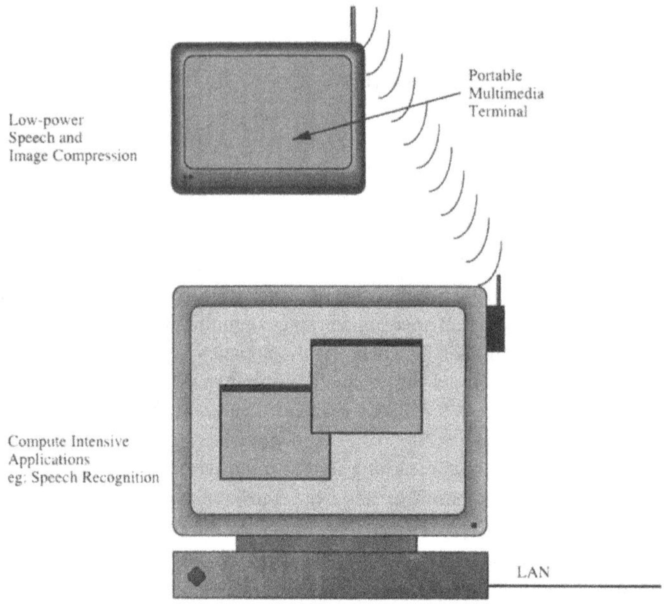

Fig. 1. Portable multimedia terminals and host network.

feature maps or Kohonen maps may be regarded as an extension to competitive learning networks. In addition to the data compression tasks mentioned above, Kohonen maps have also been successfully employed in other applications including speech recognition and robotics. Other extensions to the basic competitive learning algorithm are also discussed below.

2. Competitive Learning Algorithms and Vector Quantization

Competitive learning networks employ competition among a number of output units or neurons in the network. They may utilize a direct selection of the winning unit, or alternatively employ lateral inhibition among the units as in Fig. 2(a). They may also include the more complex topological interactions of feature maps. Current understanding of competitive learning algorithms owes much to early work by Rosenblatt [6], von der Malsburg [7,9], Fukushima [10,11], Willshaw and von der Malsburg [8], Grossberg [12,13], Amari [14], Kohonen [5,15,16] and Rumelhart and Zipser [17].

Competitive learning networks extract features from input data by clustering the positions of data vectors in the input space, according to certain distance metrics. This is discussed in detail below. In a hard competitive learning (winner-take-all) network, these cluster centers may be employed as a vector quantization [4] of the inputs, which serves as a compressed approximation of the data. Alternatively, in soft competitive learning, multiple units are excited simultaneously, creating a distributed representation or encoding of the input data. In this case, the units indirectly inhibit one another through a normalization of their outputs, which represent their individual probabilities or responsibilities for the current data point (Fig. 2(b)). There is substantial data compression in this case as well, since the number of units is limited and is less than the input dimensionality. An alternative application of such a network is as the first stage of a multilayer network, in which supervised learning is employed in the upper layer(s) (Fig. 2(c)). In this case, the use of unsupervised learning in the first (competitive) layer substantially reduces the learning time for the network, as compared to a fully-supervised learning algorithm such as back-propagation.

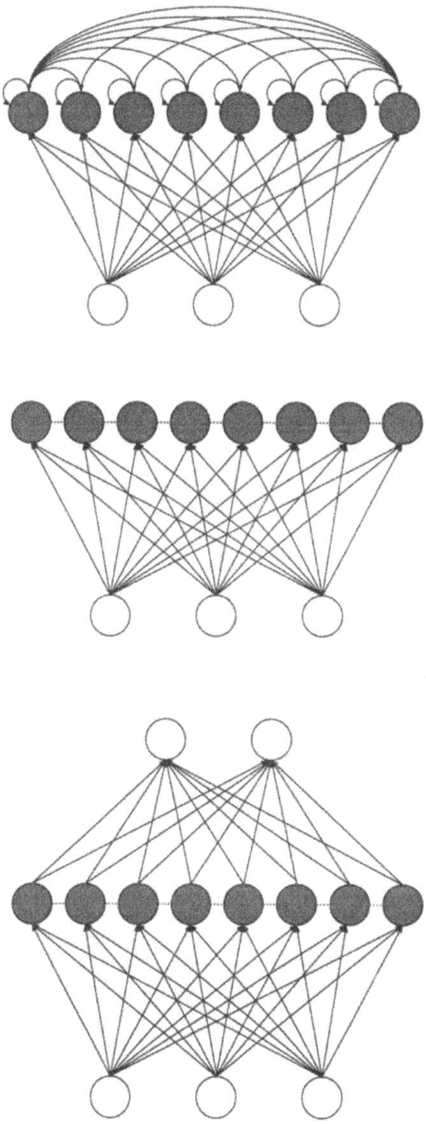

(a)

(b)

(c)

Fig. 2. Competitive learning networks (a) with explicit inhibitory connections among the competitive units, (b) with implicit inhibition via normalization as in soft competitive learning, (c) with a supervised network above which employs inputs from the competitive layer. The competitive layer can be considered to perform either data compression of data representation or the input vectors.

3. Analog or Digital VLSI Implementations of ANNs?

Analog VLSI implementations of ANNs produce compact layouts with potentially very rapid response and low power requirements, and are convenient for direct real-time interfacing to analog input sensors [18]. Digital VLSI has the advantage that it is compatible with a larger digital system and provides for reliable interchip communication, that it provides convenient access to weight values, and that it exhibits immunity to noise and to interference from other signals. However, digital arithmetic is area consuming, especially at high bit precision [19]. A compromise is to mix analog arithmetic with digital weight storage as in [20]. There are also hybrid schemes mixing analog arithmetic with digital control and communications based upon pulse streams [21]. In the following section, we discuss those properties of analog VLSI circuits, used in the implementation of artificial neural networks in general, that are relevant to the present paper. In later sections we discuss the analog and digital VLSI implementations of competitive learning algorithms specifically.

4. Analog Circuits for ANNs

Continuous-time circuits for ANNs have primarily been based on CMOS versions of transconductance amplifiers and Gilbert multipliers [18,22,23,24,25,26]. The transistors may operate in the subthreshold region of their characteristics at micropower levels, or above threshold where they exhibit speed improvements at the expense of larger but still modest power budgets. Which is preferable depends upon the application, the time scales of input and output signals, the necessity for battery operation, real-time requirements, and other factors. In addition to representing the multiplications in synapses required for learning and recall, Gilbert multipliers may also be employed as neurons with variable gain. Followed by current-voltage converters, their neural characteristics perform a good approximation to tanh functions. One input may be employed as a gain control parameter as in [24,25,26].

The other neural functions required for competitive learning networks are Gaussian activation functions for certain models and the evaluation of distortions according to Euclidean or Manhattan distance metrics, to be described below. These are in addition to the weight update calculations required by the learning computations themselves. For competitive learning networks, in the presence of lateral interactions between neurons as in Fig. 2(a), the analog circuits exhibit relaxation dynamics (settling of neural activations). However, the time scale for weight updating is generally orders of magnitude longer than that of circuit relaxation. In most cases, weight updating is performed only after the circuit has relaxed. Circuit implementations with slow but continuous weight changes during system dynamics are largely unexplored, even though this seems more biologically plausible.

In many of the original analog circuit implementations, the synaptic weights were stored in digital registers as in Fig. 3(a). Schwartz et al. later achieved 10 bits of dynamic range with capacitive weights [27]. Learning computations were generally external to the chip in early studies, with weights loaded or programmed into their respective local sites. We later reported incremental associations based on in-circuit analog Hebbian learning of capacitive weights [24,26] as in Fig. 3(b) with potential advantages for nonstationary tasks. Weights could be refreshed by periodic rehearsal on the training data. A general feature of a learning rule which scales to large networks is that it depends only on the instantaneous values of local variables (V_i and V_j and also possibly W_{ij} in Fig. 3).

4.1. Analog Weight Storage and Precision for Learning

Requirements for analog networks may include short-term, medium-term, and long-term storage methods. Short-term storage employs sample and hold techniques, medium term storage requires refresh circuitry, and long-term storage employs analog EEPROMs [28]. Horio and Nakamura [29] have comprehensively reviewed analog storage techniques for neural computation. Refresh techniques include comparison with predefined quantized levels applied as a ramp, a method with an accuracy of 7–8 bits [30]. Lee and Gulak [31] improve this refresh scheme using switched-capacitor techniques to increase the noise margin of the stored voltage. Field-programmable analog arrays [32] employing these refresh methods were used to prototype several neural circuits. Other refresh methods are reported by [33,34,35]. An alternative method is to employ periodic repetitions of the training data for refreshing the weights [24,25,26,36]. This has the advantage of adapting to temperature drifts in the circuitry and other variations in the operating environment, as well as being applicable to learning nonstationary tasks.

Murray [37] has shown that 10–20% noise added to weight updates and neural activations during analog learning accelerates learning and improves generalization. This contrasts with digital learning where as much as 12–16 bit resolution is often required, depending on the application and learning rule [38]. This may be a consequence of the LSB restriction on minimum weight update size in the digital case. Precision is low in analog learning, but the minimum update size is usually extremely small, which is beneficial late in the learning cycle. It is important to distinguish the precision required from the dynamic range required in learning updates. In analog learning circuits one can have a large dynamic range in spite of low internodal precision, since the weight updates employ local information and adapt to variations in the local circuitry performing the arithmetic computations. This is in contrast to circuits employing digital synaptic weights, in which case the minimum weight update is tied directly to the precision [39]. For example, with 4 bit weights the minimum weight update is one part in 16 and the dynamic range is limited. Stochastic weight updates may partially alleviate this problem. In [40] we have also demonstrated the distinction between random variations in the learning circuitry due to fabrication processes (which remain fixed and may be locally adapted to) and those due to noise.

A simple supervised learning algorithm for analog circuitry is the weight perturbation technique of Jabri and Flower [41]. This method measures the error gradient using the network itself, rather than calculating it as in most learning rules. A generalization of this approach, which performs parallel weight perturbations by many neurons at once, has been proposed by Alspector et al. [42]. The principle is to induce random parallel perturbations, followed by local correlations, to induce stochastic gradient descent in a (broadcast) global error measure. Zero-mean independent noise sources are employed based on LFSRs as in their Boltzmann machines [43]. Similar generalizations have been made by Flower and Jabri [44] and by Cauwenberghs [45]. A continuous-time version of this process employing noise injection and multiplicative correlation is described by Kirk et al. [46,47]. Unsupervised versions of these algorithms are also possible, since these may also perform gradient descent in known objective functions. As indicated in Fig. 2(c), unsupervised competitive learning in the lower layer of a network is often combined with supervised delta learning in the upper

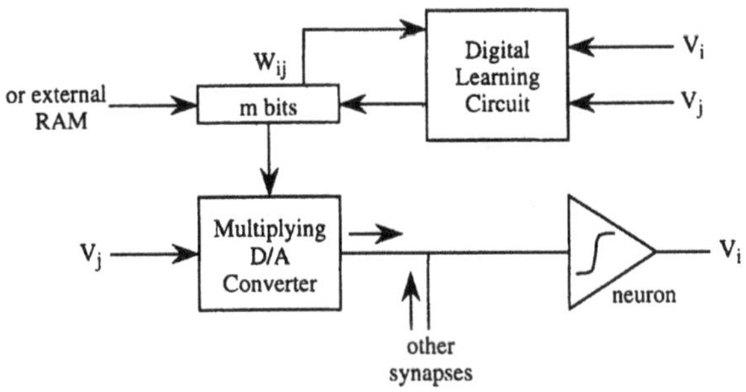

(a)

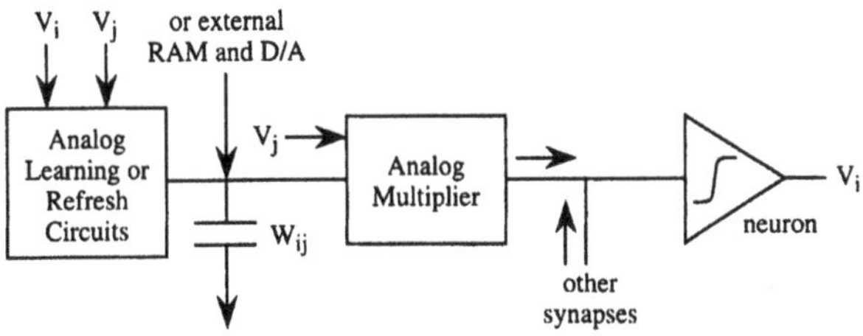

(b)

Fig. 3. Two approaches to VLSI implementation of ANNs. In the upper diagram, the synaptic weights are stored in binary registers, whereas in the lower figure the weights are represented as charge on capacitors. Both on-chip learning and the alternative external learning modes are indicated.

layer to reduce learning time from that typical of backpropagation networks.

4.2. *Analog EEPROM Weights*

Vittoz et al. [28] have identified several limitations of floating gate EEPROMs: accurate adjustment to the threshold voltage has to be performed by iterative updates, relatively large programming voltages V_p are required, and after a modification is made, a drift is observed in the threshold voltage. They have suggested a single-step programming technique involving time-dependent ac voltages on the control gate with dc voltages on the injector, with negligible voltage drift and linear dependence of the write characteristics on injector voltage (0 to 5V, for 10–100 ms programming times). The larger control signal voltages are generated externally and may be applied in parallel to all devices. To reduce V_p, they describe the use of UV light to charge floating gates, as also employed in [48]. Cauwenberghs et al. [49] describe analog CMOS floating-gate matrix-vector multiplier networks with on-chip learning, in which weights are updated under UV illumination, the intensity of which controls the learning rate. It was also demonstrated that UV-activated conductances facilitate learning on multiple time scales together with slow forgetting dynamics [50], and network models employing these devices seem biologically plausible. There have appeared several neural systems based upon the analog EEPROM array (ETANN) from Intel [51]. This chip is capable of 2GCPS in feedforward mode, and contains 10K synapses and 128 effective neurons. Multichip prototyping boards accommodate up to 8 ETANN chips, and support chip-in-the-loop training, which takes account of device variations. An 8-bit weight resolution for analog storage has been achieved using iterative programming, with on-chip generation of the programming voltage V_p.

5. On-Chip Learning

We have employed analog learning circuitry as in Figs. 4 and 5 [24,25,26] to compute weight updates in both simple incremental Hebbian learning and contrastive Hebbian learning (CHL). This method is an example of the approach shown above in Fig. 3(b).

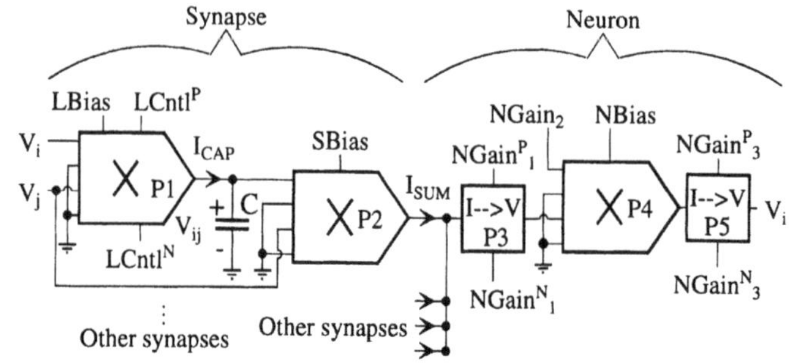

Fig. 4. Schematic diagram of neuron and synapse with on-chip Hebbian learning [24,26].

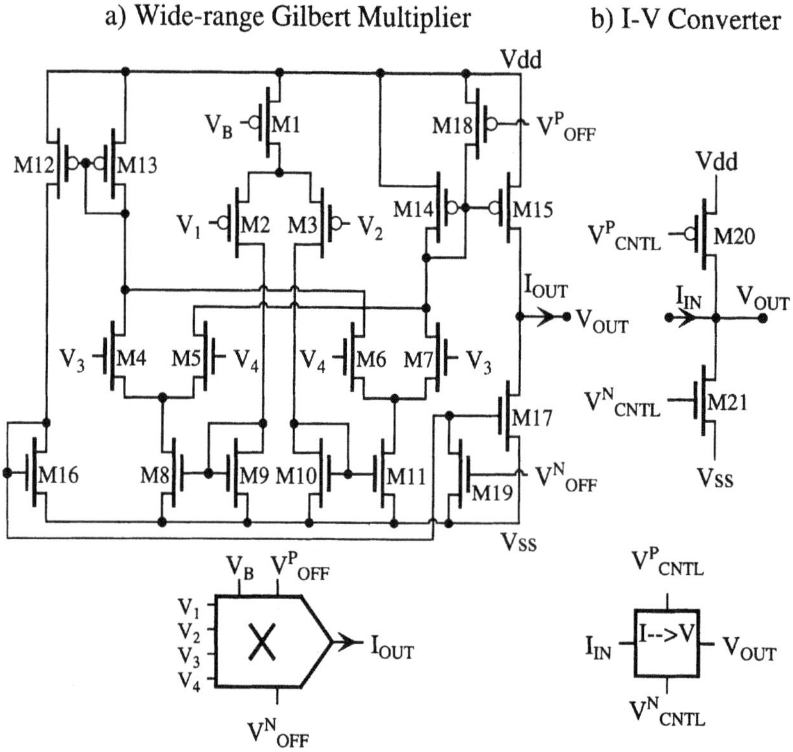

Fig. 5. Schematic diagram of typical CMOS wide range Gilbert multiplier and current-voltage converter circuits [24].

Contrastive or CHL learning rules are very similar to expressions such as Eqn. (9) below for soft competitive learning, since they employ a difference in the products of two local analog variables to update the weights. Learning is performed using currents which charge or discharge a capacitor whose analog voltage represents the weight value. Weights may also be represented by the difference in charge on two capacitors as in [27], but the ideality of our Gilbert multipliers as current sources makes this unnecessary (Fig. 6). These measured characteristics were typical of Gilbert multipliers in our $3.0\mu m$ and $1.2\mu m$ CMOS test chips, an example of which is shown in Fig. 7. The neurons are fully connected by synapses with in-situ Hebbian learning. The missing synapses on the diagonal represent the absence of self connections. These chips have been tested and found to successfully perform associative matching tasks following incremental weight updates to repeated training patterns [25]. Nonstationary tasks may be handled by repeated exposure to (slowly-varying)

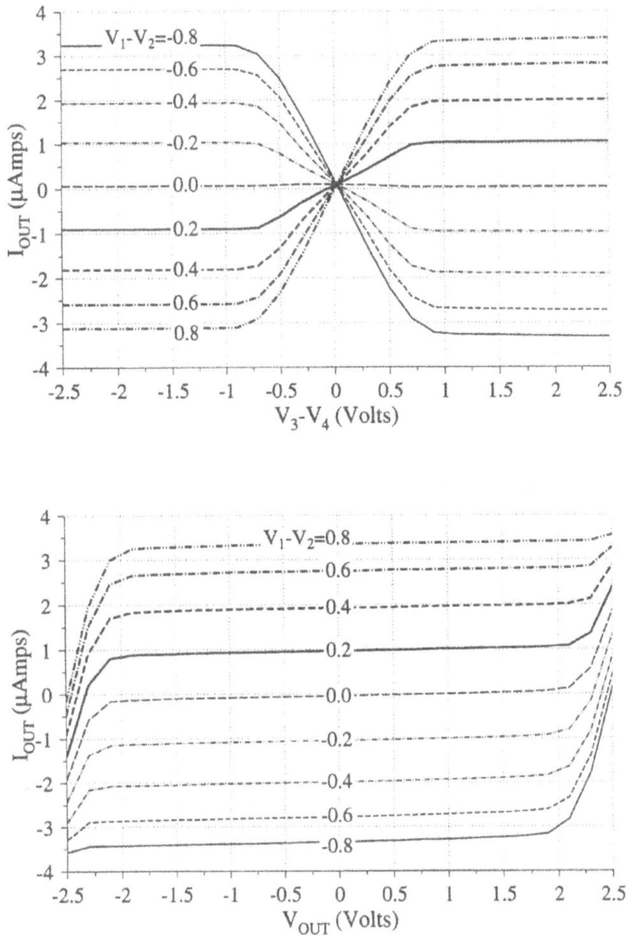

Fig. 6. Measured response of Gilbert multipliers. Upper figure shows multiplier characteristics and lower figure behaviour as a current source [24].

training data, which also compensates for temperature drifts and other environmental changes in the circuit behaviour. In addition, this precludes the need to otherwise refresh the synaptic weights. Manhattan updating was used which means that weight changes are in fixed increments governed only by the sign of the difference in Hebbian terms in the two phases, which simplifies the circuitry for computing these updates without deterioration of learning performance. A dynamic range in excess of 10 bits was observed experimentally for updates to these capacitive synaptic weights. Their precision as measured by intrachip variation is much lower, but in-situ learning makes this precision less important. Intrachip variation in device characteristics was measured to be 3–5% and interchip variation to be 30–40% in our $1.2\mu m$ CMOS chips. The effects of imperfect components in the multiplication operations cause no problems. Offsets encountered in subtracting the two Hebbian terms in CHL, or the two terms in Eqn. (9) for example, are important however and may drive the weights to extreme values. These effects may be reduced by thresholding the difference in the two terms prior to Manhattan updates, at the expense of small increases in output errors.

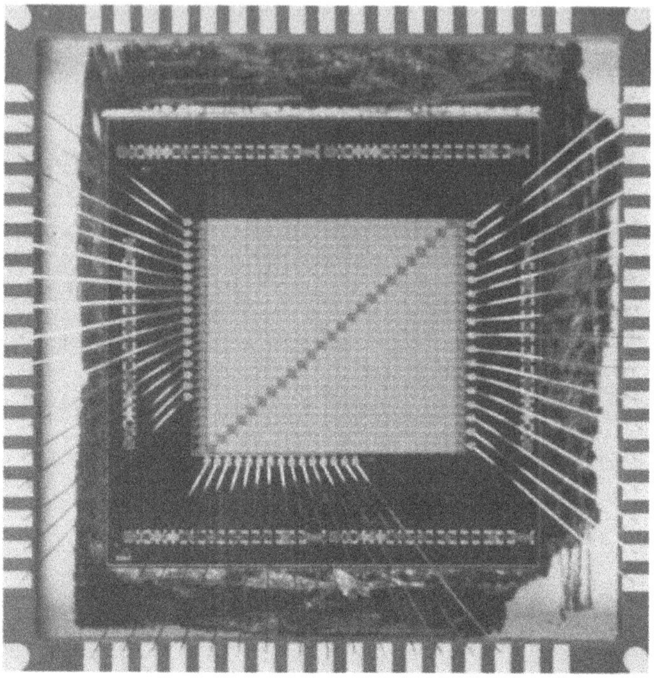

Fig. 7. Analog ANN implementation of CMOS chip with on-chip Hebbian learning in the synapses (25 neurons, 600 synapses) [26].

Arima et al. [35,36] have also reported analog learning of capacitive weights in CMOS networks. In their designs, charge pump circuits change weights in increments of 10% of their full-range values. Chips containing up to 400 neurons and 40K synapses for 1 bit neural activations and 5 bit weights perform as many as 80 GCUPS in learning mode. The large neural integration levels are a result of employing only (0, 1) values of activations and coarse learning updates. These chips may be cascaded and have been shown to learn multiple binary patterns, with weights again refreshed by periodic repetition of the training set.

One can instead employ digital weights in analog circuits as in Fig. 3(a). An advantage with digital weights is that a digital host computer has access to the weight matrix at any time during learning. Alspector and his colleagues at Bellcore have designed and fabricated several analog chips with digital on-chip learning [52,53] governed by a CHL learning rule. Weights are stored in local binary registers which drive multiplying digital-to-analog converters (MDACs) as described by Raffel et al. [54]. Digital processors local to the synapses perform the weight updates. The limited weight resolution with register based weights places a lower bound on the magnitude of the weight changes that may be learned. This limitation may be offset to some extent by stochastic weight updates. 5 bit weights (4 bits + sign bit) were employed by the Bellcore group on a chip with 32 neurons and 496 bidirectional connections [52,53]. They have implemented both stochastic and deterministic Boltzmann machines, with digital weight processors to perform learning computations. It was observed that, in the presence of a global weight decay, the limited dynamic range of 5 bit weights did not generally compromise learning. An exception was noted for large replication problems.

6. Hard Competitive Learning: Winner-Take-All Networks

Unsupervised and semi-supervised learning rules circumvent the excessive training times of supervised networks. Competitive learning (CL) models are related to clustering techniques in classical pattern recognition [12,55,16,17,56]. In hard CL, a single node is the win-

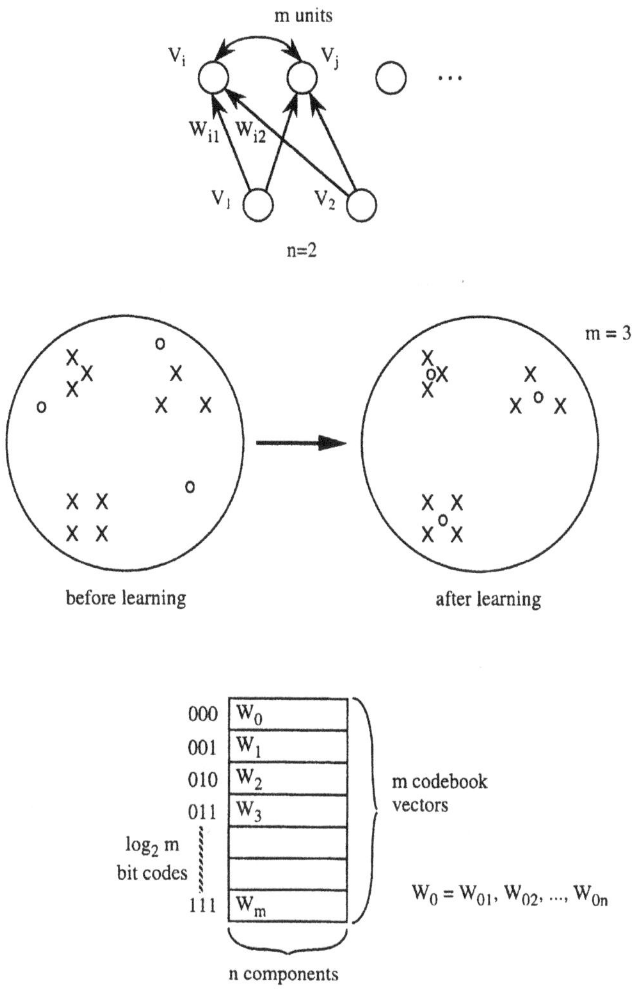

Fig. 8. Competitive learning models showing mutual interaction which is usually lateral inhibition among the neurons in the network. Figure also shows normalized input vectors together with weight vectors of the units. Also indicated is vector quantization codebook.

ner on each input pattern—a winner-take-all (WTA) network. This is employed in adaptive vector quantization (Fig. 8). WTA networks can be implemented by having inhibitory connections among all the competing nodes, i.e., lateral inhibition, and each node must also have a self-excitatory connection. The strengths of the inhibitory connections and the nonlinear activation functions of the nodes must be judiciously chosen to avoid oscillation while producing a single excited node. For each input vector, the distortion D_i for every unit is computed by determining the distance of the input vector from its weight vector. The minimum distortion may be obtained by finding the unit i whose current weight vector $\mathbf{W}_i$ exhibits the minimum Euclidean distance to the n-dimensional input vector $\mathbf{V}$.

$$D_i^2 = \sum_{j=1}^{n} (V_j - W_{ij})^2 \tag{1}$$

Once the winner is decided its weight components W_{ij} are updated according to the standard competitive learning rule [16]

$$\Delta W_{ij} = \epsilon(V_j - W_{ij}) \tag{2}$$

This learning rule (locally) minimizes the objective function

$$E = \frac{1}{2}\sum_{ij\mu} M_i^{\mu}(V_j^{\mu} - W_{ij})^2 \tag{3}$$

where M_i^{μ} is the binary cluster membership matrix element which specifies whether unit i is the winner on input pattern μ. Eqn. (2) takes on a special meaning [55] for normalized inputs and weights

$$\sum_j V_j^2 = 1 \tag{4}$$

and

$$\sum_j W_{ij}^2 = 1 \tag{5}$$

and in this case learning has, for a given node, the geometric analogy of moving weight vectors on the surface of a unit sphere into the center of gravity of all those input vectors (also on the unit sphere) for which it is the winner (Fig. 8). Also note that for normalized inputs and weights, minimizing the Euclidean distance is equivalent to maximizing the inner product, $\sum V_j W_{ij}$ since it corresponds in this case to simply obtaining the best direction for the weight vector. In other words, since

$$\sum_{j=1}^{n}(V_j - W_{ij})^2 = \sum_{j=1}^{n}(V_j^2 - 2V_jW_{ij} + W_{ij}^2) \tag{6}$$

the similarity of unit i to the input pattern may therefore in this case be computed as

$$S_i = \sum_{j=1}^{n}(V_jW_{ij}) \tag{7}$$

The similarity S_i in Eqn. (7) may also be used to compute the activation of each competing unit. The inner product in Eqn. (7) is usually easier than the Euclidean distance to compute in VLSI. If the full distortion of Eqn. (6) is required, for example if the inputs and weights are not normalized, the Manhattan or 'city-block' distance approximation to the Euclidean distance is often employed [16]. This is further discussed in a later section.

7. Frequency-Sensitive Competitive Learning

One difficulty experienced with the standard hard competitive learning algorithm as described above is that certain units may be permanently unable to win by virtue of having initial weights which were not well-matched to the data distribution. It has been proposed that a conscience mechanism [57] be applied to winning units to prevent other units from being permanently left out of the competition. One version of these algorithms is frequency-sensitive competitive learning (FSCL), in which the distortion or mismatch between an input and a weight vector increases with the frequency of selection of that unit as the winner [58]. A typical learning procedure for FSCL is to determine the distortion for each unit according to

$$D_i^2 = f_i\sum_{j=1}^{n}(V_i - W_{ij})^2 \tag{8}$$

where f_i is the frequency with which that unit has been selected as the winner on preceding input vectors in the training set.

8. Competitive Learning for Vector Quantization

The n-dimensional input data to a competitive learning network may correspond to a vector of samples from speech or image data. The weight vector of the winning unit or prototype is then an approximation, or vector quantized version, of the original sample vector. Once the competitive learning (or other method) is complete, the set of weight vectors for the various units form a codebook (Fig. 8). The approximate data may then be sent and interpreted by a receiver having a copy of the same codebook, simply by sending the address of the appropriate codevector from the table. For m units, this requires only log m bits of information and therefore performs (lossy) data compression, with a compression ratio that improves as larger errors (fewer units or codebook entries) can be tolerated. Hard competitive learning (also FSCL) is therefore one method of performing vector quantization. For improving the approximation (reducing the information loss) a reconstruction error may also be sent. In this case, soft competitive learning, discussed below, may provide a better VQ method as it better represents the data [G. E. Hinton, private communication]. A variation on standard VQ entitled learning vector quantization [5] employs some labelled training data to improve or fine-tune the clustering procedure when it is employed as a pattern classifier.

9. Circuits for Hard Competitive Learning, Mutual Inhibition, and Vector Quantization

Hard CL or WTA circuits have natural analog VLSI implementations and have been described by several groups, for example [59,60,61,62]. Also studied were local inhibition circuits based on resistive grids [18]. The basic idea of most of these circuits is mutual inhibition among competing CMOS amplifiers. O(N) interconnect is sufficient for a winner-take-all (WTA) function on N neurons. In the subthreshold region, voltages are logarithmic functions of transistor currents. It is easy to specify capacitance and current values for stable behaviour (real poles), with the time constant of the winning node several orders of magnitude smaller than that of the losing nodes [63]. Soft forms of competitive learning may also be obtained with different circuit parameters. Local WTA circuits with competition over a restricted neighbourhood is achieved by simply adding a resistive grid between the neuron connections. Neighbourhoods may be tuned by adjusting certain current ratios in the circuit [59].

Mann and Gilbert [64] describe analog chips with unary digital weights which implement an approximation to CL. Hochet et al. [30] have reported a similar implementation which stores multiple discrete levels of synaptic weights on a capacitor. The weights are refreshed without the need for full A/D or D/A conversion by employing a voltage ramp which is gated by clock signals to charge a capacitor.

Analog networks have been implemented as combinations of switched-capacitor and pulse stream networks, for example [65,66,67]. In their simplest form the synaptic weight is an analog voltage and the neural activation is a clocked pulse stream. Increasing frequency of the pulse stream reduces the effective resistance at the input to the integrator by simulating the resistor by a switched capacitor, as in switched capacitor filters. The input current is integrated and converted to a voltage which depends on the ratio of two capacitors. This voltage in turn drives a VCO to generate a pulse stream representing the next neural activation. Similar techniques have been employed in bidirectional associative memories and competitive learning circuits [68,69]. Among the virtues of this approach are the dependence of V on a capacitance ratio rather than on absolute capacitance values, and the relative economy of transistors per synapse. These methods avoid the use of synchronous clocks as in digital stochastic arithmetic systems which may alleviate interference problems. Further details and related implementations are described in [70]. Other WTA circuits are reported in [60,62,71].

10. Soft Competitive Learning and Radial Basis Functions

In soft competitive learning, all units or neurons update their weight vectors in proportion to their activation V_i

$$\Delta W_{ij} = \epsilon V_i (V_j - W_{ij}) \tag{9}$$

rather than having a single winner which updates its weight vector. Here V_i is the excitation of a given node i and V_j is a component of the input vector. Note that this rule again minimizes a known objective function E [55]. This learning rule can be thought of as a version of Hebbian learning with a particular type of weight decay.

Radial basis functions [72] implement a soft form of competitive learning in which each node is activated according to its distance from an input pattern. A common example of a distance function is

$$V_i = C\exp\left[-\frac{|\mathbf{V} - \mathbf{W}_i|^2}{2\sigma_i^2}\right] \tag{10}$$

where $\mathbf{V}$ and $\mathbf{W}_i$ are vector quantities, σ_i is the standard deviation of the gaussian activation V_i for unit i and may be a learned parameter in addition to the weight components W_{ij}.

Soft competitive learning with RBF units has a natural statistical interpretation [73,74] in terms of arriving at the best model of the input data distribution as generated by a mixture of gaussians. In this interpretation, the adaptation of each gaussian to a pattern is proportional to the likelihood that it generated that data point. Note that if this probability is proportional to

$$p(\mathbf{W}_i) = \frac{1}{\sqrt{2\pi}\sigma_i}\exp\left[-\frac{|\mathbf{V} - \mathbf{W}_i|^2}{2\sigma_i^2}\right] \tag{11}$$

then the log of this probability is given by

$$\log p(x) = C_1 - C_2|\mathbf{V} - \mathbf{W}_i|^2 \tag{12}$$

Maximizing a product of probability densities that the input data was generated by a particular mixture of gaussian distributions is the same as minimizing a sum of squared distances. In other words, by employing gaussians, least squares problems in the log probability domain are made to correspond to maximum likelihood problems [75]. Statistical procedures such as the expectation-maximization (EM) algorithm [76] can be applied to soft competitive learning by employing normalized probabilities derived from Eqn. (11) and by adapting both the mean and the variance of each gaussian for each input data point. Note that this eliminates the learning rate parameter. As with our earlier discussion concerning objective functions, with soft competitive learning we again know precisely what is being optimized.

Eqn. (11) for the neuron activation does not quite correspond to the probability of generation of the data point by gaussian unit i because it has not been normalized. Regarding the various units in the competitive layer as competing hypotheses, we can remedy this situation by dividing Eqn. (11) by the sum of the activations of all units in this layer. Note that this normalization factor provides the competition, or soft form of mutual inhibition, without the need for explicit inhibitory connections among the units. It does however demand a nonlocal computation by the neurons. In an iterative learning computation, however, it may be possible to employ a satisfactory local approximation to this normalization factor by sequentially passing information to nearest neighbours.

That soft competitive learning is an appropriate learning rule can be confirmed by observing what happens when a new data point falls approximately midway between the weight vectors of two distant units as in Fig. 9. Hard competitive learning would make the maximum weight adjustment in this case because of the large $(V_j - W_{ij})$ factor in Eqn. (2). Only the unit which happened to be marginally closer would have its weight adjusted, which would be strongly pulled toward the data point. Soft competitive learning, on the other hand, would make very small adjustments to the weight vectors of both units in this case. Neither unit would claim a large responsibility for the data point when V_i in Eqn. (10) is small for both units if the point is far from both their current weight values.

In the preceding discussion, the winning units on similar data patterns were at arbitrary positions within their layer, dependent only upon their initial weight values. Units which are winners on similar data patterns need not be near one another. In a later section we discuss means to enforce topological relationships among the units so that similar data patterns excite similar regions of the competitive layer. Often this layer is arranged in a two-dimensional sheet or map, but the layer could be one-, three-, or higher-dimensional. Attempting to learn a two-dimensional topological relation would correspond to looking for a two-dimensional subspace in the original higher-dimensional data (dimensionality reduction). This goal is similar to that of linear statistical methods such as principal components analysis [77].

11. Distortion Metrics and VLSI Computations

The topic of distortion measures was introduced using the Euclidean metric of Eqn. (1). A variety of other distortion or similarity measures have been discussed in the literature on pattern recognition and ANNs, for example [56,16,78]. A fairly straightforward generalization is the Minkowski metric in which the power of 2 in Eqn. (1) is replaced with p as in the expression

$$D_i^p = \sum_{j=1}^{n} (V_j - W_{ij})^p \tag{13}$$

A common special case of this is $p = 1$ (absolute value) which is referred to as the Manhattan or city block distance, given by

$$D_i = \sum_{j=1}^{n} |V_j - W_{ij}| \tag{14}$$

This distortion measure is computed more simply in hardware than is the Euclidean distance, since the squaring operation is avoided, and provides a reasonable approximation to it for small distances. Importantly, the Manhattan distance increases monotonically with the Euclidean distance, which is essential in a competitive network.

The next generalization of Eqn. (1) is to the Mahalanobis metric, in which the relative importance of the various input dimensions are no longer equal. This metric is given by

$$\begin{aligned} D_i^2 &= \sum_{j=1}^{n} \frac{1}{2\sigma_{ij}^2} (V_j - W_{ij})^2 \\ &= (\mathbf{V} - \mathbf{W}_i)^T \Sigma_i^{-1} (\mathbf{V} - \mathbf{W}_i) \end{aligned} \tag{15}$$

where $\mathbf{V}$ and $\mathbf{W}_i$ are vector quantities. Note that $\sigma_i = (\sigma_{i1}, \sigma_{i2}, \ldots, \sigma_{im})$ is now an m element vector quantity, and Σ_i is a $m \times m$ diagonal matrix whose m diagonal elements are $1/(2\sigma^2)$. We can regard the Mahalanobis distance as resulting from replacing the gaussian activation function, or radial basis function, for the neuron in soft competitive learning by the

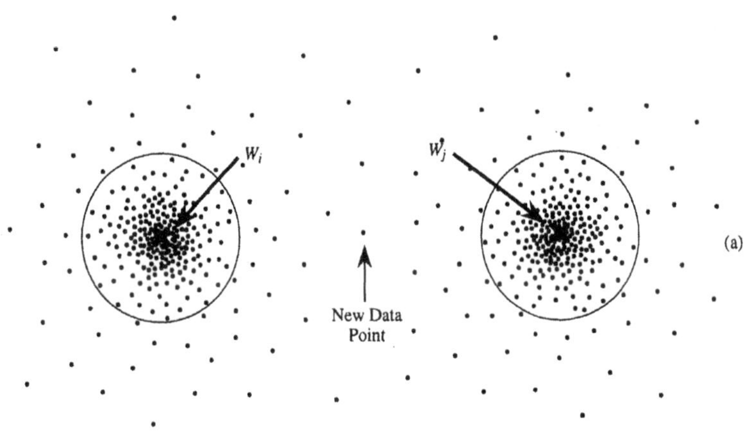

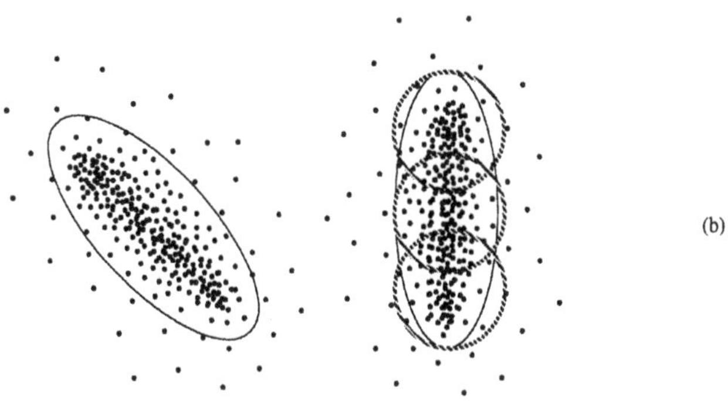

Fig. 9. Examples of input distributions and weight vectors as cluster centres (following learning). Data point half way between clusters belongs equally to both. Extended and rotated distributions may be fit with single complex radial basis functions or with multiple simple RBFs.

generalized expression

$$V_i = C\exp\left[-(\mathbf{V}-\mathbf{W}_i)^T \Sigma_i^{-1}(\mathbf{V}-\mathbf{W}_i)\right] \quad (16)$$

in which the scalar variance σ_i of Eqn. (10) has been replaced by the (diagonal) matrix Σ_i. If the off-diagonal elements of Σ_i are also non-zero (Σ_i becomes a full covariance matrix) the RBFs can also be fit to skewed data distributions as in Fig. 9. Note that this case cannot be handled by simple coordinate rotations in input preprocessing, because both skewed clusters and those aligned with the coordinate axes will exist simultaneously in general data distributions. We also point out that the input distributions could alternatively be fit to a larger number of simple RBFs (three rather than one in Fig. 9) which may in many cases be more computationally efficient. This would be the case if computing the neuron activation of Eqn. (16) were to outweigh the costs of multiple evaluations of Eqn. (10) together with the associated extra learning computations.

Since the computation of distortions given by Eqns. (13), (15) and (16) become progressively more difficult (computationally intensive) for hardware, it turns out that the simpler Eqn. (14) is typically used in implementations. The other common approach is to employ the inner product Eqn. (7) as a similarity measure, as discussed earlier in connection with normalized inputs and weights.

12. Relations to Other Unsupervised Algorithms

As mentioned above, hard competitive learning is a means of arriving at the appropriate codebook for vector quantization, by training on typical input samples. Soft competitive learning can also be used for VQ, but the approximation to an input sample following training would be reported by stating the indices of several codevectors corresponding to those units having the largest activations by that input pattern. Eqn. (2)

performed in batch mode over all input patterns corresponds to the classical k-means clustering algorithm [78,79]. Fuzzy clustering improves on k-means algorithms, and has been discussed by many authors such as [80], in which a generalization of hard competitive learning or vector quantization is proposed which may update multiple nodes in response to a single input pattern. In this respect, the fuzzy clustering method is similar to soft competitive learning as discussed above. It is argued that the prototype vectors arrived at by this process are insensitive to initialization and choice of learning coefficients. Similarly, probabilistic winner-take-all networks perform updates on units other than the winner and therefore also generalize hard competitive learning.

Adaptive resonance theory (ART) network models [81,82,83] are another generalization of CL which enforce the formation of stable categories. This is accomplished by having an abundance of node units which are only enlisted as needed. Inputs are accepted by existing nodes according to a vigilance parameter r which determines how close they must be to these nodes. One can obtain coarse or fine categorizations depending on the value of r. This approach addresses the so-called stability-plasticity dilemma, which requires that a network provide a stable response to past inputs while at the same time adapting to novel inputs. The novel inputs are handled by bringing in new units as required. This does not affect the properties of the pre-existing units, whose cluster centers are unchanged so that they may continue to properly represent their respective regions of the input data space. ART models therefore have the advantage over simple CL networks that they automatically determine the appropriate number of units. This determination must otherwise be performed as a separate phase of a CL algorithm. ART models have been extensively discussed by Carpenter and Grossberg in the literature, for example in the references above, so we will not describe the algorithms further in this paper. We do however mention VLSI implementations of ART models in the discussion below.

13. Circuits for Soft Competitive Learning and Related Algorithms

Radial basis functions, which implement soft competitive learning, lack the direct mutual inhibition of the WTA case discussed above. Watkins and Chau [84] employ analog subthreshold circuits to perform the difference, squaring and summation functions required for the distance calculations, and the evaluation of an exponentially decaying function of this distance, for each RBF neuron. Anderson et al. [85] also describe efficient analog circuits for the computation of RBFs based on computation of Gaussians to approximate quadratics. This is related to the earlier circuit of Delbruck [86], which performs a Gaussian or bump function. An alternate approach to the bump functions is to subtract two voltage-shifted tanh functions of the type shown in Fig. 6. This approach employs the circuits of Fig. 5, which are more complex than those of Delbruck, but would be very flexible in view of the independent control of the gain of each of the two tanh functions and of the relative shift.

Circuits which implement a spline approximation to nonlinear continuous functions [87] are also closely related to RBFs. These circuits learn the function $g(x)$ from a set of discrete data points x with supervision. The network applies an input at point x on a resistive grid (1-D in [87]) which spreads or diffuses this input over a localized region. Various receptive field shapes may be obtained depending on the transistor types which implement the resistors in the grid. A group of transconductance amplifier circuits in the local region become excited by input from the spreading layer, and each of these multiplies the input signal by a weight stored on its local capacitor. The function is therefore learned in approximation as a sum of basis or bump functions. A second amplifier at each grid point performs the learning according to a delta or LMS learning rule where the transconductance of this amplifier is proportional (in the subthreshold region) to the local input excitation from the spreading layer. A 1-D spline with 512 weights was accommodated in 15mm^2 for 2μm CMOS, and resulted in a weight update cycle under one microsecond corresponding to 500 MCUPS [87]. The local representation of the data implies that learning time grows linearly in the number of weights. This is unlike distributed representations in multilayer perceptrons (with sigmoids) which learn very slowly because of the large number of mutually-dependent weights.

The analog circuits based on Gilbert multipliers, described in an earlier section, directly compute the inner products of Eqn. (7) for neural activation. In addition, they are readily adapted to the soft competitive learning computation given by Eqn. (9), which is a Hebbian learning rule, together with a modified weight decay term. The Gilbert multipliers employ V_i and V_j as inputs in computing the first term in Eqn. (9), and V_i and

W_{ij} for the second term. The difference of two such terms can be directly computed by the analog circuitry in a similar manner to that demonstrated previously for contrastive Hebbian learning rules [25].

In [88], a nonlinear Hebbian learning rule replaces $V_i V_j$ by $f(V_i)\,g(V_j)$ where f and g are odd-symmetric functions. These rules may be employed in Herault-Jutten networks, which perform independent component analysis. Applications are in real-time identification tasks which separate out or identify N independent sources from N linear combinations of these which are received. This is essentially a continuous-time recursive linear adaptive filter. An example of this task is performed when the nervous system mixes information about the speed and position of body joints, and it is the task of the brain to extract the primary information. Networks employ synapses based on Gilbert multipliers modified into bump circuits [86] which implement the nonlinear rules. A subthreshold (translinear) current-mode CMOS circuit with 6 neurons and 30 synapses including on-chip learning appears in [89]. Similar work on Herault-Jutten networks was performed earlier by Vittoz and Arreguit [90].

ART models have been implemented as analog circuits by many authors, for example [91,71]. As discussed above, adaptive resonance models are another generalization of CL which enforce the formation of stable categories. The ART-2 model [81] has been implemented in analog circuits in [71]. RBFs and ART models are usually implemented (apart from learning) as feedforward rather than relaxation networks.

14. Topological Feature Extraction or Self-Organizing Maps

In early versions of feature mapping, the inhibitory (winner-take-all) connections employed in circuit implementations of hard competitive learning were instead represented by more complex lateral connections, such as the well-known Mexican hat function of Fig. 10. These types of lateral interactions are known to occur in biological systems such as mammalian vision. The prototype units form relaxation networks within their layer, with both excitatory and inhibitory connections among closely-spaced neurons. As is well-known in Hopfield networks, the existence of both $+$ and $-$ weights gives rise to frustration, in analogy with spin glasses [55]. In feature maps, correlated weight updating by neighbouring nodes can lead to topology-preserving mappings. The topology is enforced by employing a learning rule such as Eqn. (9) and by allowing the lateral connections within the prototype layer to determine the V_i term in this equation. Units near to the winning unit have their V_i values elevated through the Mexican hat interaction (Fig. 10), whereas those further away have their V_i values depressed. The networks proposed by von der Malsburg [7,8] and by Kohonen [15] are well-known examples of these maps.

In later versions of Kohonen feature maps [5,16], there are no lateral connections among the competing neurons. Instead, weight vectors for neurons in the neighbourhood of the winning node are updated as well as the weight vector of the winner itself, regardless of their excitation. This can be achieved by a learning rule such as

$$\Delta W_{ij} = \epsilon N_i (V_i - W_{ij}) \tag{17}$$

where N_i is from a neighbourhood function in which index i determines the displacement of node i from the winning node. The neighbourhood in which N_i has a significant nonzero value typically shrinks with time during the course of learning, as progressively finer topological details are ironed out. The neighbourhood function may be either a Gaussian or a step function of position, and its time dependence may be linearly or exponentially decaying, or inversely proportional to t [5]. In this case, it is the learning rule (rather than the earlier lateral interneural connections) which for these networks gives rise to the topology-preserving mappings.

15. Quantization of Weights

Kohonen's algorithm has usually been investigated assuming real values for weight variables and activations and the use of continuous or high precision arithmetic. Quantization of synaptic weight dynamics affects the convergence of learning and the ability to properly fit a given input data distribution. This influences the choice of adaptation gain and neighbourhood function (NF). A spatially-decreasing NF works far better than a rectangular neighbourhood in the presence of weight quantization with a limited bit precision [92,100].

Two-dimensional maps were generated from two-dimensional input data in discrete time in [100]. Inputs and weights were bounded without qualitatively affecting the learning behaviour. Weights were also quantized to b bits and this, in contrast, can have a major

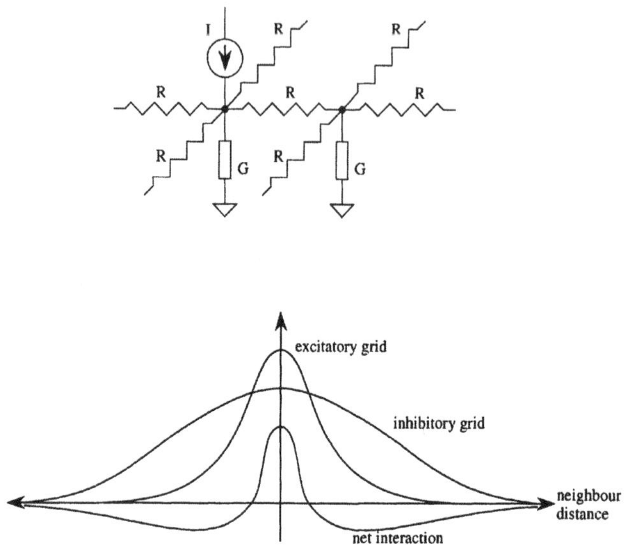

Fig. 10. Resistive grids employed in signal aggregation and filtering. In vision tasks, these networks perform Laplacian operations. With capacitive elements they detect motion of edges. A combination of two such networks can perform Mexican-hat lateral interactions in self-organizing maps [28]

effect on learning at small values of b. The minimum number of bits depends upon the number of neuron units, and the probability distribution of the input data, as b affects the asymptotic weight values following learning. For a 10×10 map and uniform data generation on a square, at least 5 bits were required due to this restriction, with mean squared error decreasing as b increases, and good performance being obtained for $b \gtrsim 7$. The minimum b also depends separately upon the rate of decrease of the adaptation gain, or learning rate $\epsilon(t)$. It is important that the decrease in the gain be slower when the bit precision is low so that the topology becomes defined before $\epsilon(t)$ drops below the minimum weight update value. This was also demonstrated in [100] for precisions in the range of 4–7 bits. The minimum b is also affected by the choice of neighbourhood function. With $\epsilon(t) = 0.55/(1 + 0.0005t)$ and a triangular NF with a range $r(t) = 1 + (6/(1 + 0.001t))$ rapid convergence to a solution with acceptable error is obtained with $b = 5$.

Spatially-decreasing NFs result in smaller errors in the final weights and learning also converges faster than with rectangular NFs. With spatially decreasing NFs, learning operates best when $\epsilon(t)$ and $N_i(t)$ in (17) are slowly decaying together, as expected. On the other hand, with rectangular NFs when $\epsilon(t)$ is slowly decaying, the rate of decay of $N_i(t)$ must increase. Otherwise, towards the end of the learning cycle, when the winning node makes no further weight updates due to the roundoff error with b bits, distant neighbours can continue to adjust their weights since their mismatch to the data is much greater. In other words, although rectangular NFs can be made to operate properly, the algorithm is less robust, with judicious choice of $\epsilon(t)$ and $N_i(t)$ being mutually dependent as well as being dependent upon b.

16. Circuits for Kohonen Networks

Analog circuits for Kohonen networks have been reported by several groups [64,92,93,94,95,96,97,98, 99,100,101]. Vittoz et al. [93] show that a Mexican-hat type of lateral interaction (with resulting activity bubble) can be created from resistance-conductance networks of the type shown in Fig. 10, which creates spatially-decreasing weighting functions. The difference between the weighting of two such RG networks, one excitatory and one inhibitory, produces the desired result. This is similar to the computations employing resistive grids used in vision and discussed below. It is however difficult to create far-reaching interactions in this way, and a better method employs voltage-dependent resistors in the network, based on transistors operating in the triode region of their characteristics [94]. Further work on this approach to activity bubbles and the estimation of Euclidean distortion metrics with high dynamic range appears in [96,101].

As discussed in an earlier section, the later method employed by Kohonen for feature maps was to directly apply neighbourhood functions during weight updating, rather than first enforcing localized neural activity via the long-range lateral inhibition of Mexican Hat-type functions. Hochet et al. [95] also suggested the use of cellular automata for generating the Kohonen neighbourhood functions (NFs). A similar nonlinear analog circuit [94] may be used to generate a NF with a conical shape in Kohonen's algorithm [100], which is a good approximation to the pyramidal shape employed in the simulations referred to above. At the same time this circuit generates the adaptation gain $\epsilon(t)$. [100] have also shown how quantization of synaptic weight dynamics in these CMOS circuits affects the convergence of learning and the ability to properly fit a given input data distribution, and how this in turn influences the choice of adaptation gain and neighbourhood function.

The recent thesis by Heim [92] discusses the Kohonen algorithm, and the use of cone-shaped learning neighbourhoods as discussed in [100]. It also addresses the computational requirements of this algorithm, and the implications for analog circuit implementations. The problems of multichip implementations, the effects of circuit inaccuracies on the performance of the algorithm, and a robust winner take all circuit for multichip implementations are described. In this study, analog voltages are employed for interchip communications. Another study from the same group has explored the use of pulse streams for interchip communications [102]. At the expense of additional voltage-frequency conversion circuits, the latter method has the advantage that dynamic range in signals may be traded against speed of operation. Another digital-analog hybrid implementation has been reported by [99] in which the synaptic weights and input signals are 8-bit binary values and the learning computations are analog. Synaptic weights are stored on capacitors in discrete voltage steps as in [95]. A 4×4 neuron chip has been fabricated in $1.6\mu m$ CMOS technology, and these chips may be cascaded to create larger Kohonen maps. Manhattan distortion metrics were employed in this study. The generation of the neighbourhood function around the winning neuron is accomplished in [92] by using a nonlinear diffusion network derived from the resistance-conductance networks explored in [18]. The on-chip learning makes use of weight storage on capacitors which are connected in the feedback paths of operational transconductance amplifiers (OTAs) circuits. A careful analysis of the interaction between leakage effects, learning rates, and the area required for the capacitors has been made, and the proposed circuits were fabricated and tested. It is difficult to establish performance predictions for large networks based either upon these studies or those reported above. Synaptic area, learning cycle time, and power consumption are comparable with those on simple competitive learning networks but there are additional requirements for the generation of neighbourhood functions in this case.

Another one-dimensional Kohonen network was reported by [98]. 12×10 synapses (12 dimensional input vectors, and 10 competitive units or prototypes) were located in $0.9mm^2$ for 190 synapses per mm^2 in $2\mu m$ CMOS. An adaptive weight memory employed switched-capacitor circuits, the activation unit used a charge comparator, and the WTA operation employed a multiport charge-sensing amplifier. Weights were again stored as charge on capacitors. Leakage in this case was tolerated by the algorithm which simply allows the losers to forget, and is claimed to be satisfactory for a sufficiently large clock frequency (above 100kHz) which determines the required arrival rate of new input vectors. Weight increments as small as 3.5 mV were possible using 1.2 pF weight capacitances.

In [97] currents represent signals and synaptic weights are stored on gate capacitances, which must be refreshed to compensate for leakage currents. Errors were explored which result from switch channel-charge injection, leakage currents, capacitive coupling, and finite output conductance. These could be alleviated through device sizing, and the use of more complex current copier circuits and memory cells. The refresh operation reads a current and writes the nearest upper reference current into the cell. This assumes the decay is less than one reference level, and requires reference current sources, effectively using n successive approximations for n bits. Their multilevel memory circuit was measured to have a resolution of 7 bits. A one-dimensional Kohonen map (extendable to two dimensions) was fabricated with on-chip learning, also using these current copier circuits. A WTA operation finds the winning node employing a Manhattan distance or distortion measure. This chip also employs a triangular neighbourhood function whose slope may be fixed by an external clock frequency, and the learning rate can also be controlled. Only 9 transistors were required per synapse plus 51 per neuron. 78% of the chip area is devoted to memory cells, consisting of 14×15 memory elements in 2.9mm^2 of 2.4μm CMOS including refresh circuitry, and the chip power dissipation was 25mW. The density would be equivalent to 5×10^4 synapses per cm^2 in 1μm CMOS. A relatively low clock frequency of 100kHz was employed due to their conservative technology.

Other distributed interactions among neurons are based on resistive grids for spatial filtering (often together with capacitive delay circuits for temporal filtering). These nets impose geometrical order on synaptic weights in addition to the topological constraints of feature maps, and have most often been used in artificial sensory processing. Recently Chua et al. [103] have described cellular neural networks for these purposes which are a form of analog cellular automata. These authors describe the analog cellular networks on hexagonal grids, with circuits based on OTAs. Harrer et al. [104] demonstrate the equivalence of these networks to resistive grids. The use of subthreshold CMOS analog circuits has been widely studied in the processing of sensory signals in audition and vision, including retinal and cochlear modelling [18]. Most of this work employs resistive grids to perform signal aggregation and filtering tasks by network relaxation. In the case of vision, photoreceptor arrays with logarithmic responses, together with arrays of horizontal resistor (HRes) circuits are integrated on chip. These chips are able to perform motion detection, figure-ground segregation, and similar tasks. Learning has so far not been a focus of this work, although recent models include short-term adaptation among other temporal effects. A number of techniques have been employed to produce resistive layers including lightly doped polysilicon lines and sheets, as well as sheets discretized in time and space using controlled diffusion of charge packets in CCDs. Most implementations however have employed grids of artificial HRes resistors based on subthreshold OTAs. These appear as resistors for small voltage, and saturate at large voltages. The resistance of these elements is controlled by an external gain input.

17. Tolerance of Competitive Learning to Analog Circuit Limitations

It has been well demonstrated empirically, for example in [59,60], that VLSI competitive learning will tolerate component imprecision and can be effective when implemented in analog circuitry for current CMOS technology. It is desirable to know exactly what degree of analog circuit variations due to fabrication and noise these circuits will tolerate, in order to anticipate their successful operation at future integration levels where these variations are expected to increase due to diminishing device and circuit dimensions. We have therefore performed a systematic study of various analog circuit limitations and their impact on competitive learning computations. The network simulations were modelled on the experimental (tanh) analog circuit characteristics of CMOS Gilbert multipliers [26] corrupted by Gaussian-distributed circuit variations and noise. Further details may be found in [105]. Soft competitive learning networks, as well as those with various degrees of lateral inhibition among the competing units, were explored in a 3-to-8 binary decoding problem. 3 analog inputs (binary signals plus noise) excited 8 competitive units, and these units attempted to learn the 8 distinct minterms without supervision. Groupings of 20 learning trials were performed with various degrees of deviation in the synaptic multiplier characteristics that are employed in the learning computations. In contrast to earlier work, the systematic effects of variation from fabrication were separated from those due to random noise, and multiplier gain variations were also studied separately from offset errors.

Fig. 11 shows the results of the non-ideal analog multiplications vs the standard deviation in the multiplier gain. The % correct learning in Table 1 refers to the number of trials in which, starting from random weights, the 8 network units after learning represented the proper 8 distinct minterms. Experimental measurements from current technology [25] place expected gain variations at approximately 10%, but the figure shows that variations as severe as 30% (0.3 standard deviation) are easily tolerated. Fig. 12 shows the impact of multiplier zero-crossing offset errors. Although the learning behaviour was very sensitive to these errors, we discovered that this sensitivity could be reduced to tolerable levels by introducing simple circuit modifications which imposed a minimum threshold for weight updates [105]. This is shown by the dotted curves in Fig. 12. In addition to gain variations and offsets, we also explored the effects of multiplicative and additive noise in the computations, as well as input noise. The competitive networks tolerated all of these noise problems extremely well. Some of these results, including synergistic effects of multiple types of variations, are

Table 1. Effect of analog hardware variations on competitive learning performance

Type of variation	Simulated variation	Correct learning	Expected variation
Multiplier gain variation	30%	90%	10%
Circuit noise	70%	91%	
Multiplier offsets	3%	15%	5%
Multiplier offsets w/threshold	5%	75%	5%
Input noise	30%	>95%	<10%
Cumulative effects	Expected	95%	

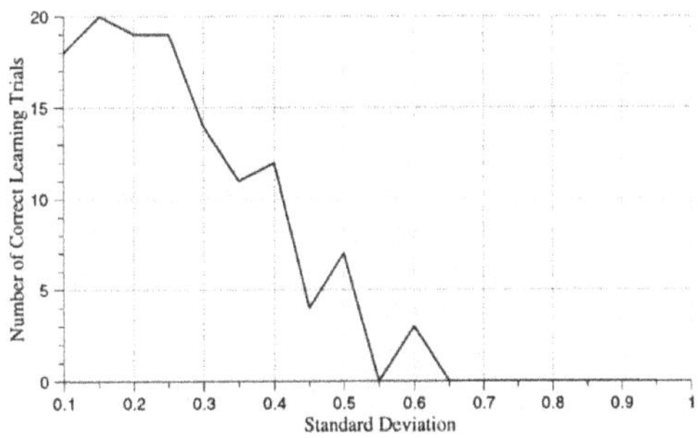

Fig. 11. Learning 3-to-8 binary decoder problem using non-ideal analog multipliers with uniform gain variations.

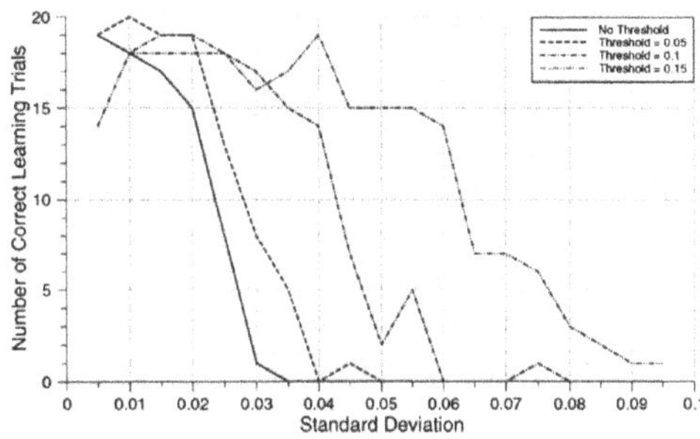

Fig. 12. Effects of offset in multiplier zero-crossing on competitive learning performance for 3-to-8 decoder problem.

summarized in Table 1. More details of these results, including dependence on learning rates, may be found in [105,106] and in the related study of variations in on-chip learning in backpropagation networks [40].

18. Applications of VLSI Competitive Learning in VQ of Speech and Images

Traditional lossy data compression methods such as code-excited linear prediction [107] provide accurate reconstruction of signals such as speaker-independent speech, but at the expense of large computational resources. These computational demands arise from extensive high precision operations such as calculation of distortion metrics during an exhaustive codebook search. It is anticipated that ANNs, on the other hand, will be able to perform parallel learning of the VQ codebook as well as codebook search with reduced arithmetic precision providing real-time adaptive vector quantization at low power in nonstationary environments. Competitive learning may ultimately result in relatively high compression ratios for a given codebook size, since the statistics of actual speech excitation data [108] will be employed in arriving at the codebook entries. Computational precision requirements generally determine energy requirements, since the number of bits of precision m demands computational resources of $O(m^2)$ in multiplication operations, either in chip area or in computation time [19]. VLSI competitive learning networks may be operated either at low digital bit precision or as analog circuitry, where the power requirements are greatly reduced [18].

Krishnamurthy et al. [58] have demonstrated the ability of ANNs to perform vector quantization of speech and images of reasonable quality and near-ideal entropy of codebook utilization, using a moderate codebook size. In their study the LPC parameters of speech signals were vector quantized using both frequency-sensitive competitive learning (FSCL) algorithms and Kohonen maps, for both speaker-dependent and speaker-independent cases. The FSCL algorithm provided the best results, comparable with those obtained using the traditional LBG algorithm for VQ, when analyzed using Itakura-Saito distortion metrics. They also examined the VQ of static images using frequency-sensitive competitive learning and obtained comparable results to LBG algorithms with respect to mean-squared error and signal-to-noise ratio.

Studies in our laboratory [109,110] have employed FSCL as well as Kohonen maps for VQ of the residual speech excitation signal (following inverse LPC and pitch prediction filters), rather than those of the LPC vocal tract filter as in [58]. The codeword index, together with the LPC and pitch parameters, are transmitted to the receiver. This approach resulted in adaptive VQ with low reconstruction error and excellent codebook entropy, with FSCL providing considerably better results than Kohonen maps. For speaker-dependent data, as the codebook size increased from 512 to 2048 entries, the entropy varied from 96.6 to 95.4% of ideal (40 ms speech segments). The segmental signal-to-noise ratio (SNR) of the reconstructed speech varied from 3.10 to 4.17 dB for FSCL, which compared favourably with code-excited linear prediction (CELP) [107] using comparable codebook sizes. The SNR is slightly better with CELP in this case, but the required computational power using CELP is also considerably larger.

In [110], for speaker-independent data, as the codebook size increases from 1024 to 8192 entries, the SNR increases from 4.96 to 8.62 dB, comparing with 7.55 dB with CELP at the largest codebook size (30 ms speech segments). Exact comparisons are difficult with CELP which among other things employs two codebooks rather than one. The bit rates or data compression ratios are also very similar to those of CELP. These results, together with informal listening tests, have confirmed that FSCL for VQ of residual signals is a viable alternative to the traditional CELP method of speech compression, particularly when computational resources (CPU power and compression time) are limited. We have also explored the dependence of SNR on the precision employed in the learning computations. These computations are able to tolerate limited precision down to 6–8 bits while maintaining acceptable quality of reconstructed speech. This shows that FSCL is a promising method of adaptive speech compression in portable applications with real-time and power constraints, as anticipated in those applications discussed in the introduction to this paper.

Adaptive VQ using ANNs which modify their network structure according to the statistics of the source signal was reported by Lee and Peterson [111]. They observed that the ANNs performed much better than the LBG algorithm for highly correlated signal sources. High compression rates (4.8kbit/s as in CELP) and respectable SNR for speech compression was also described in [112] using a hybrid method in which input speech is transformed in the frequency domain using discrete cosine transforms and the resultant vectors are quantized using a modified Kohonen algorithm. Brendan Frey in our laboratory [113] performed a similar experiment using wavelet transforms of the short term excitation. Both of these schemes are able to reduce the computation rates well below those required for speech compression using CELP methods.

Fang et al. [61] have fabricated a mixed-signal implementation of competitive learning, which they have applied to data compression by adaptive vector quantization employing a frequency-sensitive CL rule. They employ a 25 input analog chip implementing 64 codewords, and 25×64 synapse cells. The 64 competitive neurons compute Euclidean distances and a WTA circuit (with transistors in saturation mode) rapidly determines the winning unit. A digital chip maintains the input vector and codebook memory, and implements the competitive learning algorithm, D/A etc. This system handles 2 MVectors/s for 3.2 GCPS and achieves a compression ratio of 33:1.

19. Conclusions

Analog VLSI circuits for various competitive learning algorithms and networks are efficient, and lead to compact implementations. Hard and soft competitive learning, frequency-sensitive competitive learning, and topological maps are all of practical importance in data compression and representation, as well as being amenable to VLSI implementations. These circuits tolerate the analog imperfections expected from fabrication and from noise. They may provide compression algorithms suitable for applications in portable multimedia terminals with low battery power budgets. Power requirements even for highly parallel computations may be sufficiently relaxed to allow for real-time operation of the computational data compression circuitry. This would preclude any storage of the data being processed which again facilitates portability. The extent to which analog competitive learning networks are able to penetrate the multimedia applications market remains to be determined, but it is a promising direction for research.

References

1. B. Ackland, "The role of VLSI in multimedia." *IEEE J. Solid St. Ccts.* 29, pp. 381–388, Apr. 1994.
2. S. Sheng, A. Chandrakasan, and R. W. Brodersen, "A portable multimedia terminal." *IEEE Commun. Mag.* pp. 64–75, Dec. 1992.
3. S. Singhal, D. L. Gall, and C. T. Chen, "Source coding of speech and video signals." *Proc. IEEE* 78, pp. 1233–1249, July 1990.
4. R. M. Gray, "Vector quantization." *IEEE ASSP Mag.* pp. 4–29, Apr. 1984.
5. T. Kohonen, "The self-organizing map." *Proc. IEEE* 78, pp. 1464–1480, Sept. 1990.
6. F. Rosenblatt, *Principles of Neurodynamics*. Spartan: New York, 1962.
7. C. von der Malsburg, "Self organization of orientation selective cells in striate cortex." *Kybernetik* 14, pp. 85–100, 1973.
8. D. J. Willshaw and C. von der Malsburg, "How patterned neural connections can be set up by self organization." *Proc. Royal Soc. London B* 194, pp. 431–445, 1976.
9. C. von der Malsburg, "Network self-organization," in *An Introduction to Neural and Electronic Networks*, S. F. Zornetzer, J. L. Davis, and C. Lau, editors, Academic Press: San Diego, 1990.
10. K. Fukushima, "Cognitron: A self-organizing multilayered neural network." *Biol. Cybern.* 20, pp. 121–136, 1975.
11. K. Fukushima, "Neocognitron: A neural network model for a mechanism of visual pattern recognition." *IEEE Trans. Syst., Man and Cybern.* 13, pp. 826–834, 1983.
12. S. Grossberg, "Adaptive pattern classification and universal recoding: I. Parallel development and coding of neural feature detectors." *Biol. Cybernetics* 23, pp. 121–134, 1976.
13. S. Grossberg, "Competitive learning: From interactive activation to adaptive resonance." *Cognitive Science* 11, pp. 23–63, 1987.
14. S. A. Amari, "Dynamics of pattern formation in lateral-inhibition type neural fields." *Biol. Cybern.* 27, pp. 77–87, 1977.
15. T. Kohonen, "Self-organized formation of topologically correct feature maps." *Biol. Cybern.* 43, pp. 59–69, 1982.
16. T. Kohonen, *Self-organization and associative memory*. Springer-Verlag: Berlin, 1984.
17. D. E. Rumelhart and D. Zipser, "Feature discovery by competitive learning." *Cognitive Science* 9, pp. 75–112, 1985.
18. C. A. Mead, *Analog VLSI and Neural systems*. Addison Wesley: Reading, 1989.
19. J. Burr, "Digital neurochip design," in *Parallel Digital Implementations of Neural Networks*, K. W. Przytula and V. K. Prasanna, editors, Prentice Hall: Englewood Cliffs, 1993, Chap. 8, pp. 223–281.
20. P. Thiran, V. Peiris, P. Heim, and B. Hochet, "Quantization effects in digitally behaving circuit implementations of Kohonen networks." *IEEE Trans. NNs* 5, pp. 450–458, May 1994.
21. A. F. Murray, D. Del Corso, and L. Tarassenko, "Pulse-stream VLSI neural networks mixing analog and digital techniques." *IEEE Trans. on Neural Networks* 2, pp. 193–204, Mar. 1991.
22. B. Gilbert, "A precise four quadrant multiplier with sub-nanosecond response." *IEEE J. Solid State Ccts.* SC-3, p. 365, 1968.
23. F. J. Kub, K. K. Moon, I. A. Mack, and F. M. Long, "Programmable analog vector-matrix multipliers." *IEEE J. Solid St. Ccts.* 25, pp. 207–214, Feb. 1990.
24. C. R. Schneider and H. C. Card, "CMOS implementation of analog Hebbian synaptic learning circuits," in *Proc Int. Joint Conf. on Neural Networks*, Seattle, WA, 1991, Vol. 1, pp. 437–442.
25. C. R. Schneider and H. C. Card, "Analog CMOS deterministic Boltzmann circuits." *IEEE J. Solid St. Ccts.* 28, pp. 907–914, Aug. 1993.
26. H. C. Card, C. R. Schneider, and R. S. Schneider, "Learning capacitive weights in analog CMOS neural networks." *J. VLSI Sign. Proc.* 8, pp. 209–225, 1994.
27. D. B. Schwartz, R. E. Howard, and W. E. Hubbard, "A programmable neural network chip." *J. Solid St. Ccts.* 24, pp. 313–319, Apr 1989.
28. E. Vittoz, H. Oguey, M. A. Maher, O. Nys, E. Dijkstra, and M. Chevroulet, "Analog storage of adjustable synaptic weights," in *VLSI Design of Neural Networks*, U. Ramacher and U. Ruckert, editors, Kluwer: Boston, 1991, pp. 47–63.
29. Y. Horio and S. Nakamura, "Analog memories for VLSI neurocomputing," in *Artificial Neural Networks: Paradigms, Applications, and Hardware Implementations*, E. Sanchez-Sinencio and C. Lau, editors, 1992, pp. 344–363.
30. B. Hochet, V. Peiris, S. Abdo, and M. J. Declercq, "Implementation of a learning Kohonen neuron based on a new multilevel storage technique." *IEEE J. Solid St. Ccts.* 26, pp. 262–267, Mar. 1991.
31. E. K. F. Lee and P. G. Gulak, "Error correction technique for multivalued MOS memory." *Elect. Lett.* 27, pp. 1321–1323, 1991.
32. E. K. F. Lee and P. G. Gulak, "Field programmable analogue array based on MOSFET transconductors." *Elect. Lett.* 28, pp. 28–29, 1992.
33. R. Castello, D. D. Caviglia, M. Franciotta, and F. Montecchi, "Self refreshing analogue memory cell for variable synaptic weights." *Elect. Lett.* 27, pp. 1871–1873, 1991.
34. E. El-Masry and B. Maundy, "Analog ANN weight storage elements with on-chip learning capabilities using switched capacitors," in *Proc. Can. Conf. on VLSI*, Halifax, Canada, Oct. 1992, pp. 220–227.
35. Y. Arima, M. Murasaki, T. Yamada, A. Maeda, and H. Shinohara, "A refreshable analog VLSI neural network chip with 400 neurons and 40K synapses." *IEEE J. Solid St. Ccts.* 27, pp. 1854–1861, 1992.
36. Y. Arima, K. Mashiko, K. Okada, T. Yamada, A. Maeda, H. Notani, H. Kondoh, and S. Kayano, "A 336-neuron, 28K synapse, self-learning neural network chip with branch-neuron-unit architecture." *IEEE J. Solid St. Ccts.* 26, pp. 1637–1644, Nov. 1991.
37. A. F. Murray, "Analogue noise enhanced learning in neural network circuits." *Elect. Lett.* 27, pp. 1546–1548, 1991.
38. M. Hohfeld and S. E. Fahlman, "Probabilistic rounding in neural network learning with limited precision." *Neurocomputing* 4, pp. 291–299, 1992.

39. H. C. Card, "Analog circuits for relaxation networks." *Int. J. Neural Syst.* 4, pp. 359–379, Dec. 1993.
40. B. K. Dolenko and H. C. Card, "Tolerance to analog hardware of on-chip learning in backpropagation networks." *IEEE Trans. Neural Networks* 6, pp. 1045–1052, Sept. 1995.
41. M. Jabri and B. Flower, "Weight perturbation: An optimal architecture and learning technique for analog VLSI feedforward and recurrent multilayer networks." *IEEE Trans. Neural Networks* 3, pp. 154–157, 1992.
42. J. Alspector, R. Meir, B. Yuhas, A. Jayakumar, and D. Lippe, "A parallel gradient descent method for learning in analog VLSI neural networks," in *Adv. Neural Info. Proc. Syst. 5*, S. J. Hanson, J. D. Cowan, and C. L. Giles, editors, Morgan Kaufmann, 1993, pp. 836–844.
43. J. Alspector, J. W. Gannett, S. Haber, M. B. Parker, and R. Chu, "A VLSI efficient technique for generating multiple uncorrelated noise sources and its application to stochastic neural networks." *IEEE Trans. Circuits and Syst.* 38, p. 109, Jan. 1991.
44. B. Flower and M. Jabri, "Summed-weight neuron perturbation: An O(N) improvement over weight perturbation," in *Adv. Neural Info. Proc. Syst. 5*, S. J. Hanson, J. D. Cowan, and C. L. Giles, editors, Morgan Kaufmann, 1993, pp. 244–251.
45. G. Cauwenberghs, "A Fast stochastic error-descent algorithm for supervised learning and optimization," in *Proc. NIPS-5*, Denver, 1992.
46. D. B. Kirk, "Accurate and precise computation using analog VLSI with applications to computer graphics and neural networks," Ph.D thesis, Department of Computer Science, Caltech, 1993.
47. D. B. Kirk, D. Kerns, K. Fleischer, and A. H. Barr, "Analog VLSI implementation of gradient descent," in *Adv. Neural Info. Proc. Syst. 5*, S. J. Hanson, J. D. Cowan, and C. L. Giles, editors, Morgan Kaufmann, 1993, pp. 789–796.
48. D. A. Kerns, J. E. Tanner, M. A. Sivilotti, and J. Luo. "CMOS UV writeable non-volatile analog storage," in *Advanced Research in VLSI*, Santa Cruz, 1991, pp. 245–261.
49. G. Cauwenberghs, C. F. Neugebauer, and A. Yariv, "Analysis and verification of an analog VLSI incremental outer product learning system." *IEEE Trans. Neural Networks* 3, pp. 488–497, May 1992.
50. R. G. Benson and D. A. Kerns, "UV activated conductances allow for multiple time-scale learning." *IEEE Trans. Neural Networks* 4, pp. 434–440, May 1993.
51. S. Tam, M. Holler, J. Brauch, A. Pine, A. Peterson, S. Anderson, and S. Deiss, "A reconfigurable multichip analog neural network: Recognition and backpropagation training." *Proc. IJCNN 92*, Baltimore, Vol. 2, pp. 625–630, 1992.
52. J. Alspector, R. B. Allen, A. Jayakumar, T. Zeppenfeld, and R. Meir, "Relaxation networks for large supervised learning problems," in *Adv. in Neural Info. Proc. Syst. 3*, D. S. Touretzky, J. Moody, and R. Lippmann, editors, Morgan Kaufmann: San Mateo, CA, 1991, pp. 1015–1021.
53. J. Alspector, A. Jayakumar, and S. Luna, "Experimental evaluation of learning in a neural microsystem," in *Advances in Neural Info. Proc. Syst. 4*, Morgan Kaufmann: San Mateo, CA, 1992.
54. J. Raffel, J. Mann, R. Berger, A. Soares and S. Gilbert, "A generic architecture for wafer scale neuromorphic systems," in *Proc. IEEE Int. Conf. Neural Networks* III, 1987, pp. 501–513.
55. J. Hertz, A. Krogh, and R. G. Palmer, *Introduction to the Theory of Neural Computation*, Addison-Wesley: Menlo Park, CA, 1991.
56. R. O. Duda and P. E. Hart, *Pattern Recognition and Scene Analysis*, John Wiley: New York, 1973.
57. D. DeSieno, "Adding a conscience to competitive learning," *Proc. IEEE Int. Conf. Neural Networks*, Vol. 1, pp. 117–124, San Diego, 1988.
58. A. K. Krishnamurthy, S. C. Ahalt, D. E. Melton, and P. Chen, "Neural networks for vector quantization of speech and images." *IEEE J. Sel. Areas in Commun.* 8, pp. 1449–1457, Oct. 1990.
59. J. Lazzaro, S. Ryckebusch, M. A. Mahowald, and C. A. Mead, "Winner-take-all networks of O(N) complexity," in *Adv. in Neural Info. Proc. Syst. 1*, D. S. Touretzky, editor, Morgan Kaufmann: San Mateo, CA, 1989, pp. 703–711.
60. J. Choi and B. J. Sheu, "A high precision VLSI winner take all circuit for self-organizing neural networks." *IEEE J. Solid St. Ccts.* 28, pp. 576–584, May 1993.
61. W. C. Fang, B. J. Sheu, O. T. C. Chen, and J. Choi, "A VLSI neural processor for image data compression using self-organizing networks." *IEEE Trans. Neural Networks* 3, pp. 506–518, May 1992.
62. B. J. Sheu and J. Choi, *Neural Information Processing and VLSI*, Kluwer: Boston, 1995.
63. J. L. Wyatt and D. L. Standley, "Criteria for robust stability in a class of lateral inhibition networks coupled through resistive grids." *Neural Computation* 1, pp. 58–67, 1989.
64. J. R. Mann and S. Gilbert, "An analog self-organizing neural network chip," in *Advances in Neural Information Processing Systems 1*, D.S. Touretzky, editor, Morgan Kaufmann: San Mateo, CA, 1990, pp. 739–747.
65. Y. P. Tsividis and D. Anastassiou, "Switched capacitor neural networks." *Elect. Lett.* 23, pp. 958–959, 1987.
66. M. Brownlow, L. Tarassenko, A. F. Murray, A. Hamilton, S. Han and H. M. Reekie, "Pulse firing neural chips for hundreds of neurons," in *Advances in Neural Information Processing Systems 2*, D. S. Touretzky, editor, Morgan Kaufmann: San Mateo, CA, 1989, pp. 785–792.
67. A. Rodriguez-Vazquez, R. Dominguez-Castro, A. Rueda, J. L. Huertas, and E. Sanchez-Sinencio, "Nonlinear switched capacitor neural networks for optimization problems." *IEEE Trans. Ccts. Systems* 37, pp. 384–398, Mar. 1990.
68. B. Maundy and E. El-Masry, "A switched capacitor bidirectional associative memory." *IEEE Trans. Ccts. and Systems* 37, pp. 1568–1572, 1990.
69. B. Maundy and E. El-Masry, "A self organizing switched capacitor neural network." *IEEE Trans. Ccts. and Systems* 38, pp. 1556–1563, 1991.
70. H. C. Card and C. R. Schneider, "Analog CMOS neural circuits — In situ learning." *Int. J. Neural Syst.* 3, pp. 103–124, 1992.
71. R. Linares-Barranco, E. Sanchez-Sinencio, A. Rodriguez-Vazquez, and J. L. Huertas, "A modular T-mode design approach for analog neural network hardware implementations." *IEEE J. Solid St. Ccts.* 27, pp. 701–713, 1992.
72. J. Moody and C. J. Darken, "Fast learning in networks of locally tuned processing units." *Neural Computation* 1, pp. 281–294, 1989.
73. S. J. Nowlan, "Soft competitive adaptation: Neural network algorithms based on fitting statistical mixtures," Ph.D thesis, School of Computer Science, Carnegie-Mellon University, 1991.

74. G. E. Hinton and S. J. Nowlan, "The bootstrap Widrow-Hoff rule as a cluster formation algorithm." *Neural Computation* 2, pp. 355-362, 1990.
75. G. E. Hinton, *Neural Networks for Industry*. Information Technology Research Center, University of Toronto, 1991.
76. A. P. Dempster, N. M. Laird, and D. B. Rubin, "Maximum likelihood from incomplete data via the EM algorithm." *J. Royl Sta. Soc.* 39, pp. 1–38, 1977.
77. E. Oja, "Neural networks, principal components, and subspaces." *Int. J. Neural Systems* 1, pp. 61–68, 1989.
78. J. T. Tou and R. C. Gonzalez, *Pattern Recognition Principles*. Addison-Wesley: Reading, 1974.
79. J. MacQueen, "Some methods for classification and analysis of multivariate data," in *Proc. 5th Berkeley Symp. on Prob. and Statistics*, University of California Press, 1967.
80. N. R. Pal, J. C. Bezdek, and E. C. K. Tsao, "Generalized clustering networks and Kohonen's self-organizing scheme." *IEEE Trans. Neural Networks* 4, pp. 549–557, July 1993.
81. G. A. Carpenter and S. Grossberg, "A massively parallel architecture for a self-organizing neural pattern recognition machine." *Computer Vision, Graphics and Image Processing* 37, pp. 54–115, 1987.
82. G. A. Carpenter and S. Grossberg, "The ART of adaptive pattern recognition by a self-organizing neural network." *IEEE Computer*, pp. 77–88, Mar. 1988.
83. G. A. Carpenter and S. Grossberg, "A self-organizing neural network for supervised learning, recognition, and prediction." *IEEE Commun. Mag.* pp. 38–49, Sept. 1992.
84. S. S. Watkins and P. M. Chau, "A radial basis function neurocomputer implemented with analog VLSI circuits," *Proc. IJCNN 92*, Baltimore, 1992, Vol. 2, pp. 607–612.
85. J. Anderson, J. C. Platt, and D. B. Kirk, "An analog VLSI chip for radial basis functions," in *Adv. Neural Info. Proc. Syst. 5*, S. J. Hanson, J. D. Cowan, and C. L. Giles, editors, Morgan Kaufmann, 1993, pp. 765–772.
86. T. Delbruck, "Bump circuits for computing similarity and dissimilarity of analog voltages," in *Proc. Int. Joint Conf. on Neural Networks*, Seattle, July 1991, Vol. 1, pp. 475–479.
87. D. B. Schwartz and V. K. Samalam, "An analog VLSI splining circuit," in *Advances in Neural Information Processing Systems 3*, D. S. Touretzky, J. Moody, and R. Lippmann, editors, Morgan Kaufmann: San Mateo, CA, Apr. 1991, pp. 1008–1014.
88. M. H. Cohen and A. C. Andreou, "MOS circuit for nonlinear Hebbian learning." *Elect. Lett.* 28, pp. 591–593, 1992.
89. M. H. Cohen and A. G. Andreou, "Current mode subthreshold MOS implementation of the Herault-Jutten autoadaptive network." *IEEE J. Solid St. Ccts.* 27, pp. 714–723, 1992.
90. E. Vittoz and X. Arreguit, "CMOS integration of Herault-Jutten cells for separation of sources," in *Analog VLSI Implementation of Neural Systems*, C. A. Mead and M. Ismail, editors, Kluwer: Boston, 1989, pp. 57–83.
91. S. W. Tsay and R. W. Newcomb, "VLSI implementation of ART1 memories." *IEEE Trans. Neural Networks* 2, pp. 214–221, 1991.
92. P. Heim, "CMOS analogue VLSI implementation of a Kohonen map," D.Sc.T. Thesis, Ecole Polytechnique Federale de Lausanne, Switzerland, 1993.
93. E. Vittoz, E. Sorouchyari, P. Heim, X. Arreguit, and F. Krummenacher, "Analog VLSI implementation of a Kohonen map," in *Proc. Journees D'Electronique 1989*, Lausanne, Oct. 1989, pp. 292–301.
94. P. Heim, B. Hochet, and E. Vittoz, "Generation of learning neighbourhood in Kohonen feature maps by means of simple nonlinear network." *Elect. Lett.* 27, pp. 275–277, 1991.
95. B. Hochet, V. Peiris, S. Abdo, and M. J. Declercq, "Implementation of a learning Kohonen neuron based on a new multilevel storage technique." *IEEE J. Solid St. Ccts.* 26, pp. 262–267, Mar. 1991.
96. O. Landolt, E. Vittoz, and P. Heim, "CMOS self-biased Euclidean distance computing circuit with high dynamic range." *Elect. Lett.* 28, pp. 352–354, 1992.
97. D. Macq, M. Verleysen, P. Jespers, and J. D. Legat, "Analog implementation of a Kohonen map with on-chip learning." *IEEE Trans. Neural Networks* 4, pp. 456–461, May 1993.
98. Y. He and U. Cilingiroglu, "A charge-based on-chip adaptation Kohonen neural network." *IEEE Trans. Neural Networks* 4, pp. 462–469, May 1993.
99. V. Peiris, B. Hochet, and M. Declercq, "Implementation of a fully parallel Kohonen map: A mixed analog digital approach." *Proc. ICNN* 4, pp. 2064–2069, 1994.
100. P. Thiran, V. Peiris, P. Heim, and B. Hochet, "Quantization effects in digitally behaving circuit implementations of Kohonen networks." *IEEE Trans. NNs* 5, pp. 450–458, May 1994.
101. O. Landolt, "An analog CMOS implementation of a Kohonen network with learning capability," in *VLSI for Neural Networks and Artificial Intelligence*, J. G. Delgado-Frias and W. R. Moore, editors, Plenum Press: New York, 1994, pp. 25–34.
102. A. Mortara and E. Vittoz, "A communication architecture tailored for analog VLSI artificial neural networks: Intrinsic performance and limitations." *IEEE Trans. Neural Networks* 5, pp. 549–466, May 1994.
103. B. E. Shi and L. O. Chua, "Resistive grid image filtering: Input/output analysis via the CNN framework." *IEEE Trans. Ccts. and Syst. I: Fund. Th. and Appl.* 39, pp. 531–548, 1992.
104. H. Harrer, J. A. Nossek, and R. Stelzl, "An analog implementation of discrete time cellular neural networks." *IEEE Trans. Neural Networks* 3, pp. 466–476, May 1992.
105. H. C. Card, D. K. McNeill, C. R. Schneider and R. S. Schneider, "The impact of VLSI fabrication on neural learning," in *Proc. IEEE ISCAS*, Seattle, 1995, Vol. 2, pp. 985–988.
106. D. K. McNeill and H. C. Card, "Analog hardware tolerance of soft competitive learning," in *Proc. IEEE ICNN*, Orlando, 1994, Vol. 4, pp. 2004–2008.
107. M. R. Schroeder and B. S. Atal, "Code-excited linear prediction (CELP); High quality speech at very low bit rates," in *Proc. IEEE Int. Conf. Acoust., Speech, and Signal Processing*, 1985, pp. 937–940.
108. Y. M. Cheng and D. O'Shaughnessy, "On 450–600 b/s natural sounding speech coding." *IEEE Trans. on Speech and Audio Processing* 1, pp. 207–220, Apr. 1993.
109. S. Kamarsu, "Neural code-excited linear prediction for low power adaptive speech coding," MSc. thesis, Dept. of Electrical and Computer Engineering, University of Manitoba, 1995.
110. S. Kamarsu and H. C. Card, "Limited-precision unsupervised learning algorithms for speech coding," in *Proc. INNS World Congress on Neural Networks*, Washington, D.C., July 1995, Vol. 1, pp. 128–131.
111. T. C. Lee and A. M. Peterson, "Adaptive vector quantization using a self-development neural network." *IEEE J. Sel. Areas in Commun.* 8, pp. 1458–1471, Oct. 1990.
112. L. V. Veleva and R. K. Kunchev, "Adaptive speech coding with DCT and neural net vector quantization." *Elect. Lett.* 29, pp. 704–705, Apr. 1993.

113. B. Frey, "Code and wavelet excited linear prediction of speech," MSc. thesis, Dept. of Electrical and Computer Engineering, University of Manitoba, 1994.

Howard C. Card obtained his Ph.D degree at the University of Manchester in 1971. He has held academic appointments at Manchester, Manitoba, Waterloo and Columbia Universities, and has spent a sabbatical year at Oxford. He has been a consultant at IBM Watson Research Center and an in-house instructor at Bell Labs. His early research was on device physics and modelling, his later work on parallel VLSI computations. He has been an Athlone Fellow and a Mullard Fellow. He was chairman of the 1984 Gordon Conference on MIS Systems and the 1987 Canadian Conference on VLSI, as well as serving on the program committees for many other conferences. He has received several teaching awards including the Stanton Award for Excellence in Teaching and the UMFA-UTS Award. He has also received several research awards including the NSERC E.W.R. Steacie Memorial Fellowship, the Sigma Xi Senior Scientist Award, the ITAC-NSERC Award for Academic Excellence, the Ross Medal of IEEE Canada, the Rh Institute Award for Multidisciplinary Research, and Fellow of the IEEE. He has taught 15 different courses, supervised more than 30 graduate students, and produced over 200 publications. He has given many invited lectures in Canada, Europe and the U.S.A. His current research interest is in VLSI and software systems employing artificial neural learning, and their applications in adaptive user interfaces, multimedia signal processing, and educational toys. He is also interested more generally in the relationship between biology, physics, and computation.

Dean K. McNeill is a doctoral candidate in the Department of Electrical and Computer Engineering at the University of Manitoba. He is currently conducting research into the learning characteristics of competitive artificial neural network algorithms with particular emphasis on the issues surrounding their implementation in VLSI hardware and autonomous mobile robotics. He has previously obtained his M.Sc. degree in 1993, and B.Sc. degree in 1990, both from the University of Manitoba.

photo
not available
at time of
print

Christian R. Schneider received the B.Sc. (EE), M.Sc. and Ph.D. degrees from the University of Manitoba, Winnipeg, Canada in 1983, 1986, and 1991. His current research interests include analog VLSI neural network implementations, parallel processing, computer-aided design, and computer graphics. He is presently employed by Niche Technology, Winnipeg, Canada.

Analog Integrated Circuits and Signal Processing, 15, 315–327 (1998)

Design of Neural Networks Based on Wave-Parallel Computing Technique

YASUSHI YUMINAKA[1], YOSHISATO SASAKI[1], TAKAFUMI AOKI[2] AND TATSUO HIGUCHI[2]

[1]*Department of Electronic Engineering, Faculty of Engineering, Gunma University, Kiryu 376, Japan*
[2]*Department of System Information Sciences, Graduate School of Information Sciences, Tohoku University, Sendai 980-77, Japan*

yuminaka@el.gunma-u.ac.jp

Received July 2, 1996; Accepted December 16, 1996

Abstract. A Wave-Parallel Computing (WPC) technique is proposed to address the interconnection problem in massively-parallel analog hardware required for implementing neural networks. The fundamental concepts are the frequency multiplexing of signals on a single line, and their wave-parallel processing without decomposition. This paper discusses the realization of a Hopfield-type massively-connected neural network, and shows that the WPC-based network exhibits much lower topological complexity compared with the original network. We also investigate the possible implementation of WPC using the present MOS technology, and discuss the performance evaluation in terms of the degree of multiplexing and the processing speed.

Key Words: Analog Implementations of Neural Networks, Multiple-Valued Logic, Set-Valued Logic, Frequency Multiplexing

1. Introduction

The miniaturization of pattern dimensions and the resultant increase in the number of processing modules are the directions of the future highly parallel VLSI. As the number of modules increases, the resultant increase in the number of interconnections among modules becomes critical to overall system performance and cost in terms of chip area, delay time and power consumption. All these severe limitations originate from the fact that a single lead transmits only one bit information (0 or 1) at the same time, and that any sophisticated functions of VLSI systems are realized solely depending on a very primitive switching function of a transistor. Such communication crisis has created challenging tasks in the design of new paradigm computing architectures.

One promising candidate for breaking through this difficulty is no doubt the Multiple-Valued Logic(MVL) ICs [1]. For instance, four-valued signaling increases the information-carrying capacity of each connection, and hence reduces the number of total interconnections. Furthermore, a four-valued computing module executes two independent binary logic functions simultaneously, as if two binary logic modules were multiplexed into the four-valued module. However, if we realize more efficient "multiplex computing"—r-multiplex computing, 2^r-valued signaling is required. At the state-of-the-art technology, it seems to be difficult to handle such very many (e.g., $2^{10} = 1024$) values with reasonable stability.

The authors have proposed a new multiplex computing scheme called *wave-parallel computing* (WPC) technique for addressing problems caused by interconnection complexity [2]–[4]. The underlying concept is the frequency multiplexing of signals, that is, all the logical information is coded into the variety of frequency components carried by amplitude-modulated waves. This new computing scheme employs frequency space as an extra degree of freedom, a new dimension for parallel processing.

This paper is the first attempt to explore the potential of WPC architectures for realizing neural networks consisting of fully connected analog neurons. In VLSI implementations of neural networks of N neurons, $O(N^2)$ interconnectivity is required for the massive-interconnections, which would translate to $O(N^3)$ chip area [5]. Although the frequency-multiplexed signal representation reduces the number of interconnections, an appropriate demultiplexing circuitry may be required at each neuron. Clearly, this leads to the increase of hardware complexity of the entire network. Hence, the key to success in realizing WPC-based neural networks lies in finding an efficient way for processing frequency-multiplexed signals directly without separating them into individual wires. In this paper, we present *wave-parallel multiplication* technique to perform weighted-sum operations directly on frequency-multiplexed analog signals. The most attractive feature of this technique is that it combines the frequency discrimination function and the weighted-sum operation into a single circuit. This makes it possible to reduce the number of signal lines and synaptic connections in a Hopfield-type fully connected network by a factor of $1/n$, with n frequency components.

This paper is organized as follows. Section 2 introduces the principle of wave-parallel computing technique. Section 3 describes the design of neural networks on the basis of WPC concept. In this section, we propose a wave-parallel multiplication technique for neural networks in which both the computation and communication are performed simultaneously within a same circuitry. Section 4 deals with the implementation issue of WPC circuits, and discusses the performance evaluation in terms of the degree of multiplexing and the processing speed. Also, the optical implementation of the proposed network is discussed to demonstrate the future possibility of WPC architectures.

2. Concept of Wave-Parallel Computing for Neural Networks

For the real-world applications in the next generation, parallel processing architectures with very low latency will become essential. One of the central issues in their design relates to the realization of densely-connected global interconnections over which many processing modules communicate. Furthermore, intelligent information processing inherently entails many *functional* modules capable of both computation (arithmetic operation, e.g., weighted-sum operation) and communication.

Motivated by these requirements, we have proposed the concept of Wave-Parallel Computing (WPC) in which frequency multiplexing of amplitude-modulated signals are used for both communication and computation [2]–[4]. WPC lessens the number of interconnections needed within and between VLSI chips, and there is no sacrifice in processing concurrency. Consider a system consisting of n processing elements PE_1, PE_2, ..., PE_n, which produce analog/digital output values $v_1, v_2, \ldots, v_n$, respectively. In WPC, the output signal of v_i is represented by an amplitude-modulated signal of n frequency components $f_1, f_2, \ldots, f_n$ as

$$V(t) = v_1 \cos(2\pi f_1 t) + v_2 \cos(2\pi f_2 t) + \cdots + v_n \cos(2\pi f_n t), \quad (1)$$

where each f_i is a carrier frequency for multiplex data transmission. For simplicity, we assume that sufficient number of frequency components is available in our system, that is, n_n (the number of PEs) $= n_f$ (the number of frequencies) $= n$. But, the following discussion can be extended to more general case where $n_f < n_n$.

The analog signals thus represented can be transmitted simultaneously through a single lead without interference. It is noteworthy that the following discussion is also applicable to other multiplexing schemes, such as optical wavelength-division multiplexing [6].

The major drawback of this representation is that an appropriate demultiplexing circuitry may be required at each receiving node to separate the different signals $v_1, v_2, \ldots, v_n$ before processing. This leads to the increase of hardware complexity of the entire network. For area-efficient implementation, each PE should have capability of processing the multiplexed signals directly without separating them into individual discrete wires.

Addressing the above problem, we shall present *wave-parallel multiplication* technique, which enables us to combine the frequency discrimination function and the weighted-sum operation. In this technique, we use a weighted multi-frequency signal such that

$$W(t) = w_1 \cos(2\pi f_1 t) + w_2 \cos(2\pi f_2 t) + \cdots + w_n \cos(2\pi f_n t), \quad (2)$$

where $(w_1, w_2, \ldots, w_n)$ is an arbitrary weight vector. Multiplying $V(t)$ by $W(t)$ using an analog multiplier, we have

$$\begin{aligned} 2W(t)V(t) = {} & w_1 v_1 + w_2 v_2 + \cdots + w_n v_n \\ & + w_1 v_1 \cos(2\pi(2f_1)t) + \cdots \\ & + w_i v_i \cos(2\pi(2f_i)t) + \cdots \\ & + w_n v_n \cos(2\pi(2f_n)t) \\ & + (w_1 v_2 + w_2 v_1)(\cos(2\pi(f_1 - f_2)t) \\ & + \cos(2\pi(f_1 + f_2)t)) + \cdots \\ & + (w_i v_j + w_j v_i)(\cos(2\pi(f_i - f_j)t) \\ & + \cos(2\pi(f_i + f_j)t)) + \cdots \\ & + (w_{n-1} v_n + w_n v_{n-1}) \\ & \times (\cos(2\pi(f_{n-1} - f_n)t) \\ & + \cos(2\pi(f_{n-1} + f_n)t)). \end{aligned} \quad (3)$$

By eliminating high-frequency components using a simple lowpass filter as depicted in Fig. 1, the output yields

$$LPF(2W(t)V(t)) = w_1v_1 + w_2v_2 + \cdots + w_nv_n, \quad (4)$$

where $LPF()$ denotes lowpass filtering operation. If $n = 1$, the above equation means well-known coherent detection technique for amplitude-modulation (AM) transmission. It is important to note that the general form of Eq. (4) is the inner product of the two vectors $(v_1, v_2, \ldots, v_n)$ and $(w_1, w_2, \ldots, w_n)$. This enables us to achieve the weighted-sum operation together with demodulation, which is particularly useful for intelligent information processing, e.g., neural computing applications. Moreover, the wave-parallel multiplier has advantages that the circuit cost does not depend on the number of frequency components, and higher-degree of multiplexing is expected due to the inherent properties of coherent detection.

WPC architectures are classified by their mode of operation in the PE, that is, digital mode and analog mode operation. In digital mode operation, each v_i is limited to 1 or 0, namely, binary signal. We have already proposed a unified design method, based on set-valued logic concept [7]–[8], for digital mode operation, in which many binary modules are multiplexed into a single WPC module [2]–[4]. To further enhance the versatility of WPC architectures, following section discusses the application to neural network employing analog mode operation.

3. Design of WPC-Based Neural Networks

We discuss the application of analog-mode WPC architectures to neural networks (NNs) as an example of densely connected architectures. In a Hopfield-type network of n neurons, the activity of the i-th neuron is described as

$$u_i = \sum_{j=1}^{n} w_{ij}v_j + b_i, \quad (5)$$

$$v_i = g(u_i) \qquad (i = 1, 2, \ldots, n), \quad (6)$$

where u_i, v_i and b_i are the potential level, the output and the external bias of the i-th neuron, respectively; w_{ij} is the synaptic weight from the j-th to the i-th neuron; $g(\cdot)$ denotes a sigmoid-shaped function.

Figure 2(a) shows the simplified implementation of a Hopfield-type neural network in the conventional architecture for $n = 4$. It is apparent that the central issue in realizing neural paradigms is how to implement complicated interconnections with matrix-vector multiplications.

Using wave-parallel multiplication technique, Eqs. (5) and (6) can be rewritten as

$$u_i = LPF(2W_i(t)V(t)) + b_i, \quad (7)$$

$$v_i = g(u_i), \quad (8)$$

where we represent the weight vector $(w_{i1}, \ldots, w_{in})$ and the output vector $(v_1, \ldots, v_n)$ in the multiplexed forms:

$$W_i(t) = \sum_{j=1}^{n} w_{ij}\cos(2\pi f_j t), \quad (9)$$

$$V(t) = \sum_{j=1}^{n} v_j\cos(2\pi f_j t). \quad (10)$$

Figure 2(b) shows the corresponding network configuration for $n = 4$. The wave-parallel multipliers, each consisting of an analog multiplier and a lowpass filter, perform matrix-vector multiplication without separating the multiplexed signals into individual wires. In general, the number of neurons n_n may be greater than the number of available frequency components n_f, that is, $n_n > n_f$. In this case, we have to partition the given neurons into $\lceil n_n/n_f \rceil$ groups (block neurons), each consisting of n_f neurons at most as illustrated in Fig. 3. Then, the activities of the neurons within a same group are multiplexed with n_f frequencies, and a network of reduced size is employed to provide connectivity among the $\lceil n_n/n_f \rceil$ groups.

When we intend to divide an original network into m block neurons, $(m - 1)n$ additional wave-parallel multipliers are required. Hence, we must determine the adequate block size which minimizes the total circuit cost in terms of multipliers and wiring complexity. Further study on the systematic design method for such block neuron is required.

To confirm the principle of WPC-based NN, we design a 4-bit neural-based A/D converter. We set the synaptic weights w_{ij} and the biases b_i asymmetrically to avoid the local minimum as described in Ref. [9]. As a result, we have

$$W_i(t) = \sum_{i<j\leq 4} (-2^{j-1})\cos(2\pi f_j t), \quad (11)$$

$$b_i = X - 2^{i-1} + 0.5 \qquad (i = 1, 2, 3, 4), \quad (12)$$

where X is an analog input voltage. The digital values $v_1, v_2, v_3, v_4 \in [0, 1]$ are computed to be the bi-

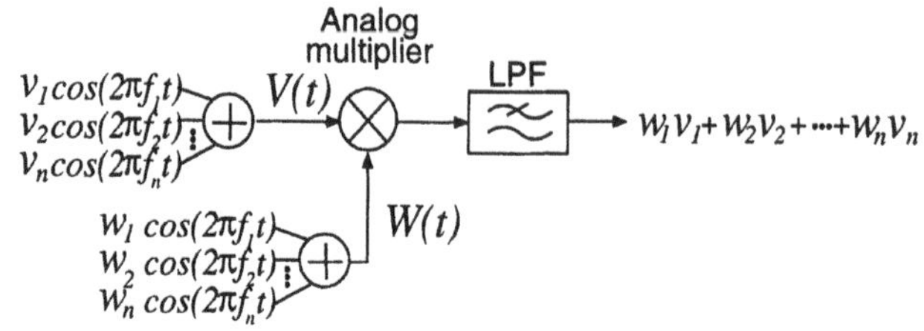

Fig. 1. Schematic diagram of the wave-parallel multiplier.

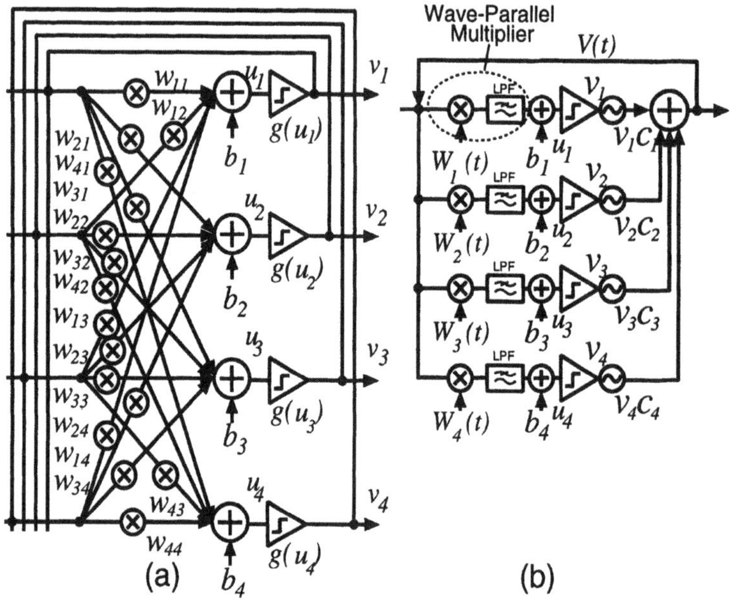

Fig. 2. Model of neural networks for $n = 4$: (a) Conventional realization; (b) WPC-based realization, where $v_j c_j = v_j cos(2\pi f_j t)$.

nary representation of X, that is, $X \simeq \sum_{i=1}^{4} 2^{i-1} v_i$. The i-th neuron employs the weighted multi-carrier signal $W_i(t)$ to perform matrix-vector multiplication. For instance, by multiplying $V(t)$ by $W_2(t)$ $(= -4\cos(2\pi f_3 t) - 8\cos(2\pi f_4 t))$ and eliminating high-frequency components, the matrix-vector multiplication for u_2 is achieved. We can obtain the output v_2 by applying threshold operation g to u_2 with threshold voltage 0V. The outputs v_1, v_2, v_3, v_4 thus obtained are frequency-multiplexed into $V(t)$, which is fed back to the neurons again (see Fig. 2(b)).

Figure 4(a) depicts the transient responses of each neuron's activities u_1, u_2, u_3, u_4 and the multiplexed output $V(t)$, where $X = 10$, $f_1 = 8MHz$, $f_2 = 12MHz$, $f_3 = 20MHz$ and $f_4 = 40MHz$. The SPICE2 simulation shows that the digital representation of 10—$v_1, v_2, v_3, v_4 = 0, 1, 0, 1$—can be obtained under the threshold 0V. (The waveform of each u_i shows the potential level before threshold operation.) Also, Fig. 4(b) demonstrates the correct network operation as an A/D converter changing input value from 0 to 15.

In general, a conventional n-neuron network requires a two-dimensional array of n^2 multipliers [5]. In the WPC architecture, in contrast, n multipliers perform the equivalent computation. Furthermore, the number of interconnections for the feedback loop can be reduced from n to 1 if sufficient number of frequencies

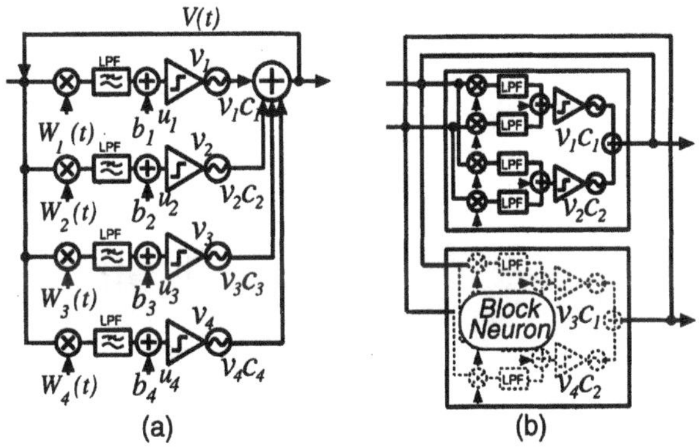

Fig. 3. WPC-based neural networks: (a) for $n_n = n_f = 4$, (b) for $n_n = 4 > n_f = 2$. In this case, an original network is divided into two block neurons.

is available. Thus, two-dimensional reduction in area complexity is expected in a fully-connected network.

The proposed architecture can also be implemented with optical devices, where intensity-modulated optical signals are employed instead of amplitude-modulated voltage signals [6]. Fig. 5(a) shows an experimental system consisting of light emitting diodes (LEDs), photodiodes, a feedback mirror and other electronic components. Each neuron generates its output as an optical signal modulated in its unique frequency. The optical signals thus generated are multiplexed through free space, and are fed back to the neurons. Each neuron converts the intensity-modulated optical signal into an amplitude-modulated voltage signal to perform wave-parallel computation. Fig. 5(b) shows the operating waveforms for 3-bit A/D conversion. The network can thus be implemented with no inter-neuron interconnections. In the future, it might be possible to integrate the opto-electronic components, such as photodetectors and light emitters, together with neuron circuits on a same substrate.

4. Realization and Evaluation of WPC circuits

4.1. Implementation issue

Implementing wave-parallel computing circuits requires analog adders, analog multipliers, lowpass filters and multi-frequency sources. Our goal is to realize these functions compactly at the present state of MOS technology.

• *Wave-parallel multiplier* The wave-parallel multiplier can be realized through an analog multiplier and a lowpass filter. The well-known four-quadrant gilbert multiplier is capable of performing wide-bandwidth operation approximately a couple of 100MHz with compact circuits [10].

Lowpass filters, on the other hand, usually require long RC time constant for low frequency operation. From the view point of integrated circuits, capacitors and resistors are not efficiently realized because they are area hungry devices. To cope with this issue, we have proposed the use of Transistor-only filters [11], which utilize the distributed RC characteristic of a MOS transistor, for compact implementation [2]–[4]. This frequency-selective operation of MOS transistors offers high-density integration of lowpass filters required for the wave-parallel multiplier.

Figure 6 shows the circuit diagram of a wave-parallel multiplier consisting of an analog multiplier and a Tr-only lowpass filter. Also, the multiplier core was actually fabricated in a 10-μm nMOS process as shown in the microphotograph. Because we adopt the relatively lower impurity concentration ($3.0 \times 10^{15}/cm^3$) wafer, the total harmonic distortion was less than 1% for $|V_1 - V_2| < 1V$, and the $-3dB$ bandwidth was measured about $6MHz$.

•*Multi-frequency source* Another key element is a multi-frequency source. Due to the properties of coherent detection, it is important to synchronize the carrier frequency between a transmitter and a receiver. This

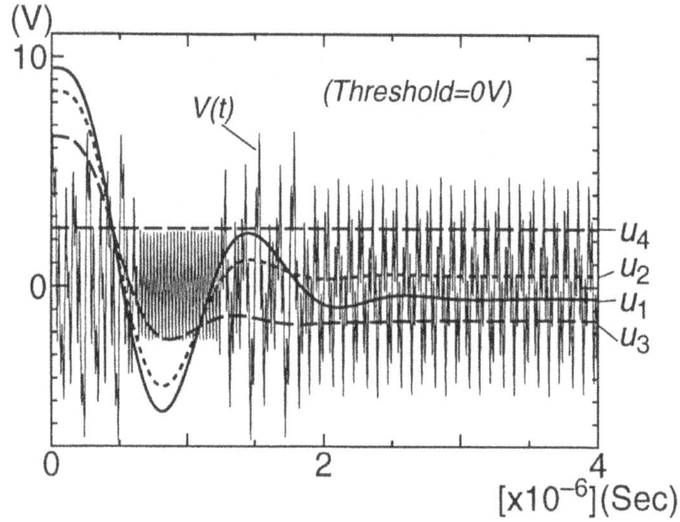

(a)

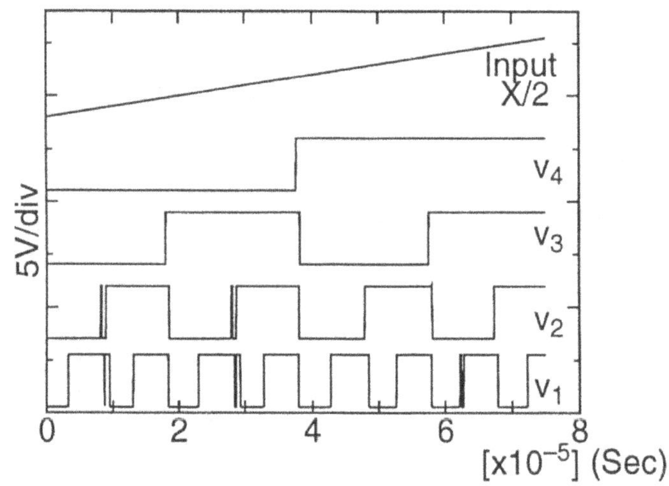

(b)

Fig. 4. (a) Transient response of the 4-bit neural-based A/D converter for input $X = 10$. In this simulation, the multiplier model of SPICE2, the 4th-order Butterworth lowpass filter and the comparator model LM311 were used. (b) Operating waveforms of the 4-bit neural-based A/D converter.

synchronization issue might be a major drawback of WPC circuits.

We discuss the effect of frequency and phase errors in the coherent detection for the highest frequency f_n. Let the modulated signal be $x_n \cos(2\pi f_n t)$ and the local carrier be $y_n \cos(2\pi(f_n + \Delta f)t + \phi)$. The local carrier frequency/phase errors in this case are Δf and ϕ, respectively. The product of the modulated signal

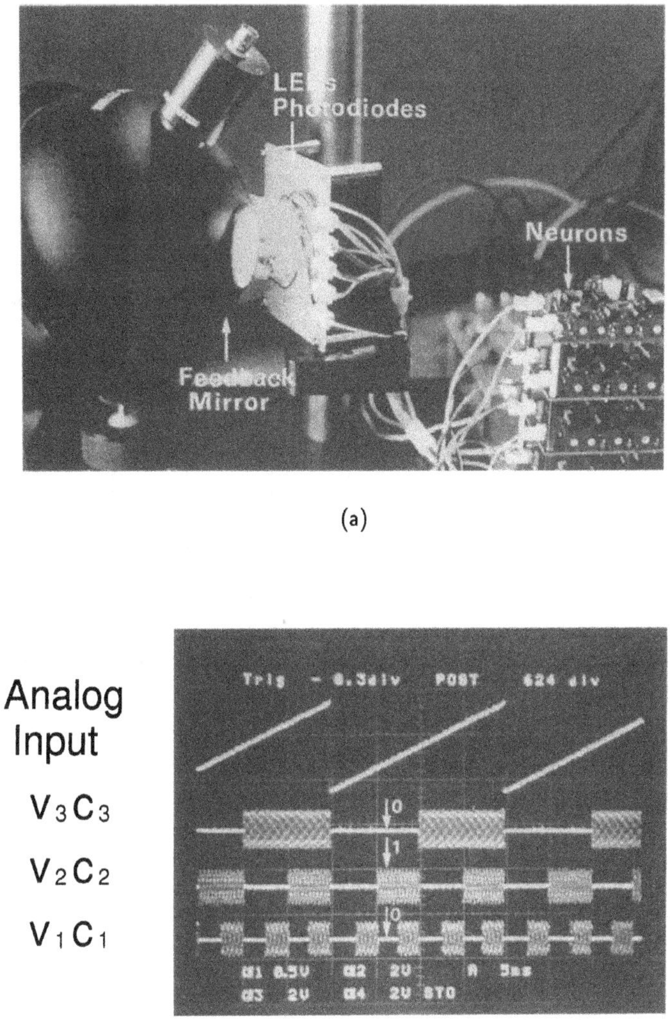

(a)

(b)

Fig. 5. (a) Experimental implementation of the WPC-based neural network using optical devices. (b) Operating waveforms of the 3-bit neural-based A/D converter using optical devices, where $v_j c_j = v_j cos(2\pi f_j t)$. (Optical intensity is converted into voltage.)

and the local carrier is given by

$$2x_n \cos(2\pi f_n t) y_n \cos(2\pi(f_n + \Delta f)t + \phi)$$
$$= x_n y_n \cos(2\pi \Delta f t + \phi)$$
$$+ \underbrace{x_n y_n \cos(2\pi(2f_n + \Delta f)t + \phi).}_{\text{Eliminated by LPF}} \tag{13}$$

The output of the filter given by $x_n y_n \cos(2\pi \Delta f t + \phi)$ is thus distorted compared with the ideal output $x_n y_n$.

This time-varying attenuation is a serious problem;

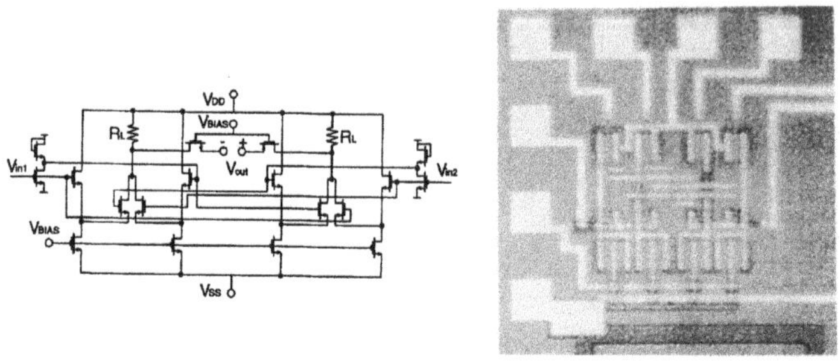

Fig. 6. Wave-parallel multiplier: Circuit diagram and microphotograph of the fabricated chip.

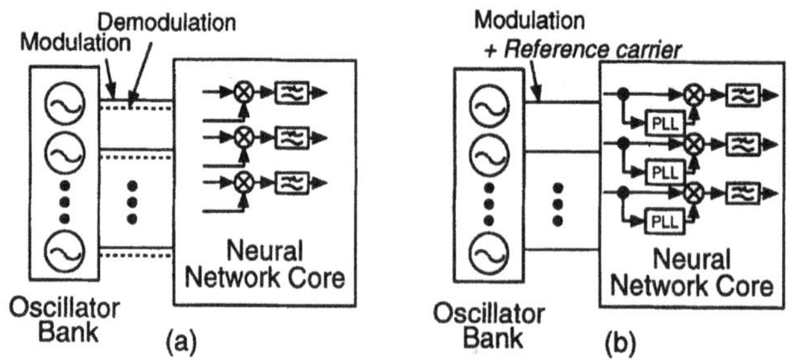

Fig. 7. Multi-frequency generation methods: (a) Supply from outside, (b) Internal generation using Phase-Locked Loop (PLL) circuits

hence the local oscillator should be properly synchronized. There are mainly two methods of generating synchronized multi-frequency signals. The direct way is that the multi-frequency signals for modulation and demodulation are supplied from the same frequency sources (oscillator bank) as shown in Fig. 7(a). Although we can obtain well-synchronized signals using this method, additional wires are required.

Figure 7(b) shows the another approach. In this case, a reference carrier may be transmitted, and superimposed on the modulation signal, which may be extracted at the receiver with a Phase-Locked Loop (PLL) to generate multiple carrier frequencies. A digital ring oscillator in a usual PLL, however, generates square waves rather than sinusoidal waves. Fortunately, the coherent detection technique can also be applied to these square-wave carriers. On the basis of this property, the WPC-based A/D converter treated before can be realized using square-wave carrier. Figure 8 indicates the SPICE2 simulation which demonstrates the successful use of square-wave carrier. Hence, we might employ square-wave carriers which can be effectively handled in digital circuits. In any case, of course, we should consider the frequency assignment carefully to avoid the harmonic components of the square waves. Since discussion on such issue as a whole is beyond the scope of this brief paper, the further discussion on realizing a multi-wave generator will be reported in the another paper.

4.2. Performance Analysis

We shall analyze a general relationship between the degree of multiplexing and the processing speed of WPC, which holds in arbitrary devices. As has been seen, multiplexing of n kinds of signals enables us to reduce the input/output lines by $1/n$. There exists, however, trade-off between the degree of multiplexing and the cycle time (processing speed). A cycle time is defined as an interval between the sequence of signals applied to an input. Hence, for the purpose of improving the performance of WPC circuits with regard to the processing speed, it is important to determine the adequate degree of multiplexing and the cycle time which meet the device and intended applications.

First, we assign frequency components to individual signals by dividing frequency bandwidth ($0 < f_i <$

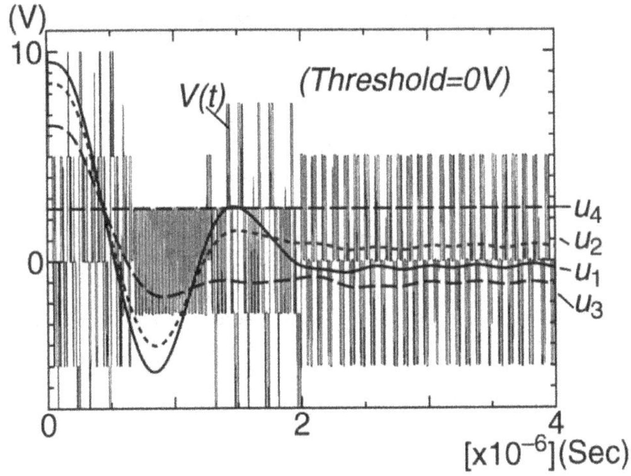

Fig. 8. Transient response of the 4-bit neural-based A/D converter for input $X = 10$ using a square-wave carrier.

f_{max}) into equal interval α as $0 < f_1 = f_n - (n-1)\alpha < f_2 = f_n - (n-2)\alpha < \cdots < f_n = f_{max}$, where f_{max} is the available highest frequency of the device. The cycle time of WPC circuits is determined by the frequency assignment. Thus, the problem treated here is to determine the optimum frequency assignment which minimizes the cycle time. To analyze this relationship, in the following, we try to express the coherent detection procedure using Fourier series.

For simplicity, we assume that WPC system can be modeled as shown in Fig. 9 consisting of modulation part and demodulation part. Furthermore, we assume that the baseband signal $V(t)$ (which is digitalized in a time domain) is approximated by the square wave of frequency $\frac{f_s}{2}$ ($f_s = \frac{1}{T_c}$, where Tc is a cycle time) as shown in Fig. 9(a).

By using 3-rd order approximation, the product of a baseband signal $V(t)$ and a carrier signal $\cos(2\pi f_1 t)$ can be represented by

$$V(t)\cos(2\pi f_1 t) = 2x_1\left(\frac{1}{2} + \frac{2}{\pi}\left(\sum_{k=1}^{3}\frac{1}{2k-1}\cos\left(2\pi\frac{(2k-1)f_s}{2}t\right)\right)\right) \cdot \cos(2\pi f_1 t)$$
$$= x_1\cos(2\pi f_1 t) + \frac{2x_1}{\pi}\left(\sum_{k=1}^{3}\frac{1}{2k-1}\left(\cos\left(2\pi\left(f_1 + \frac{(2k-1)f_s}{2}\right)t\right) + \cos\left(2\pi\left(f_1 - \frac{(2k-1)f_s}{2}\right)t\right)\right)\right) \quad (14)$$

The modulation of ideal *continuous* baseband signal $V(t) = x_1$ is simply represented by $x_1\cos(2\pi f_1 t)$. On the other hand, using WPC data representation, a baseband signal is time-limited by T_c, resulting in having complex spectra located at frequency $\frac{(2k-1)f_s}{2}$ above and below of the frequency f_1 as shown in Fig. 9(b). (In this simulation, we set the values $f_1 = 100\ KHz$ and $f_s = 10\ KHz$.)

By multiplying modulated signal (Eq. (14)) by same frequency component f_1, demodulation is achieved. Figure 9(c) shows the waveform and the spectrum of coherent detection output before lowpass filtering. The original baseband spectrum is retrieved as follows

$$V(t)\cos(2\pi f_1 t)2\cos(2\pi f_1 t)$$
$$= 2x_1\left(\frac{1}{2} + \frac{2}{\pi}\left(\sum_{k=1}^{3}\frac{1}{2k-1}\cos\left(2\pi\frac{(2k-1)f_s}{2}t\right)\right)\right) \quad (15)$$
$$+ x_1\cos(2\pi(2f_1)t) + \frac{2x_1}{\pi}\left(\sum_{k=1}^{3}\frac{1}{2k-1}\left(\cos\left(2\pi\left(2f_1 + \frac{(2k-1)f_s}{2}\right)t\right) + \cos\left(2\pi\left(2f_1 - \frac{(2k-1)f_s}{2}\right)t\right)\right)\right) \quad (16)$$

The spectra of Eq. (15) are the components of a baseband signal $V(t)$, and the other spectra are high-

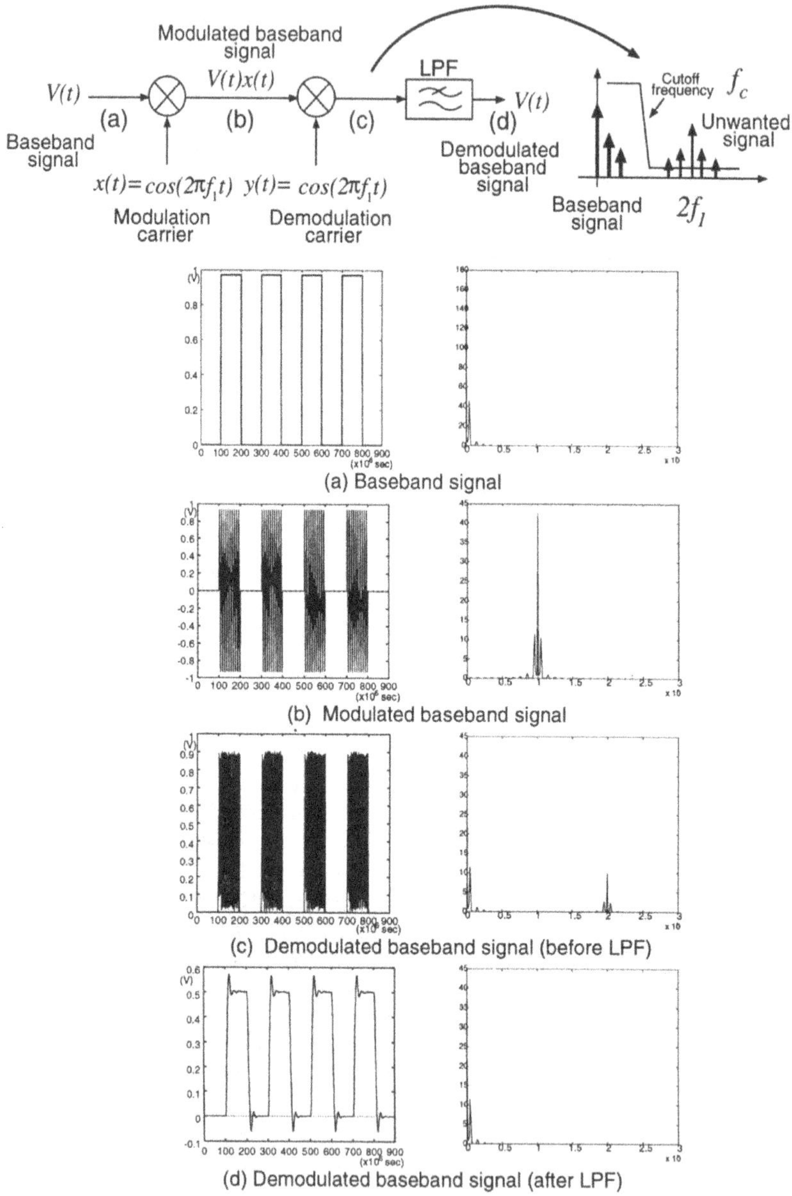

Fig. 9. Waveforms and Spectra of coherent detection.

frequency components to be eliminated by a lowpass filter (Fig. 9(d)). The lowpass filter is required to separate the highest frequency component ($\frac{5f_s}{2}$) of a baseband signal and the lowest frequency component ($2f_1 - \frac{5f_s}{2}$) of an unwanted signal. Thus, as the condition of cutoff frequency f_c, we have

$$\frac{5f_s}{2} < f_c < 2f_1 - \frac{5f_s}{2} \tag{17}$$

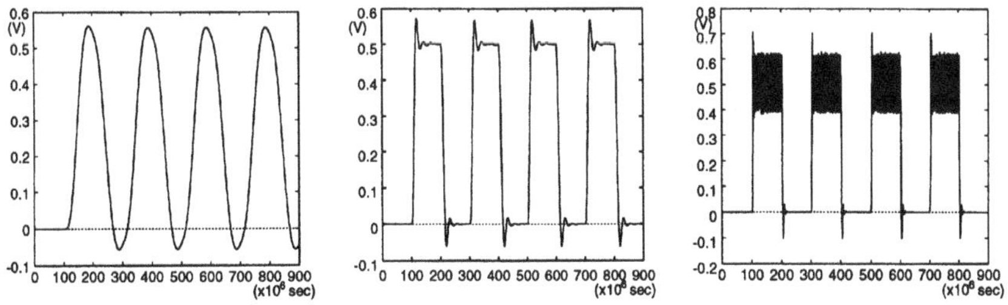

Fig. 10. Spectra of coherent detection output after lowpass-filtering ($f_c = 10KHz$, $f_c = 45KHz$, $f_c = 200KHz$).

In this simulation, the range of f_c is calculated to be $25KHz < f_c < 175KHz$. Figure 10 shows the waveforms and spectra of lowpass filter output for $f_c = 10KHz$, $f_c = 45KHz$, $f_c = 200KHz$. Low f_c also eliminates the side lobe of a baseband signal itself resulting in the deterioration of the baseband signal. On the other hand, high f_c passes the unwanted frequency components.

Next, we generalize the above discussion into the coherent detection of n frequency components. In this case, the interval between frequencies is given by

$$f_i - f_{i-1} = \alpha = \frac{2}{2n-1} f_{max}. \tag{18}$$

Using the similar procedure of Eq. (16), we have following relation

$$\begin{aligned} \frac{af_s}{2} &< f_c < f_i - f_{i-1} - \frac{af_s}{2} \\ &= \frac{2}{2n-1} f_{max} - \frac{af_s}{2}. \end{aligned} \tag{19}$$

where a is the highest order of harmonics which approximates the square wave of baseband signal ($a = 1, 3, 5, \ldots, (2n-1)$). Hence, the processing speed f_s is estimated as

$$f_s < \frac{2}{a(2n-1)} f_{max} \tag{20}$$

Figure 11 shows the relationship between the degree of multiplexing and the processing speed in the case $f_{max} = 100MHz$.

Because WPC architectures have relatively low data rates, they are not meant to compete with conventional VLSI systems capable of executing fast iterative arithmetic computation. Instead, their advantage is massive and natural parallelism. They offer a new parallel processing architecture that is not restricted to the conventional interconnected structure. Consequently, parallelism at the level of entire architecture might provide potential advantages, even if individual processing modules are somewhat larger and slower than binary gates.

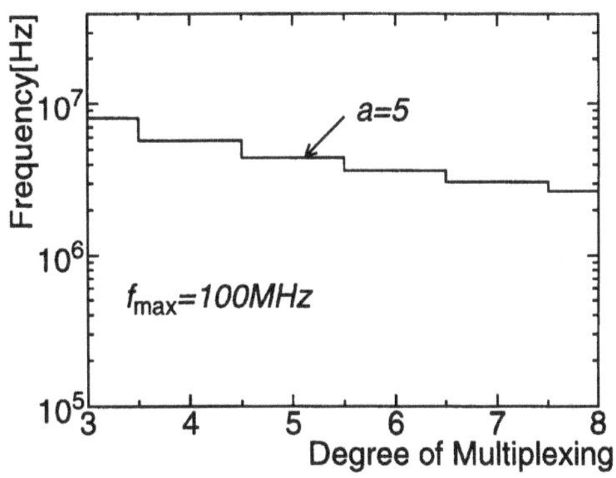

Fig. 11. Relationship between the degree of multiplexing and the processing speed.

5. Conclusion

We have proposed wave-parallel computing technique for neural networks. The purpose of this paper is to explore the possible means by which WPC concept might contribute to the solution of interconnection problems in neural networks. Computing with multiplexable information carriers offers the possibility to reduce the overall complexity of such parallel computing architecture without loss of its processing concurrency.

Further investigations are in progress to evaluate our system in terms of chip area, power dissipation and harmonic distortion in detail. Of course, WPC is not a fully established technology but which will show considerable promise for overcoming communication crisis in the next generation.

References

1. K. C. Smith and P. G. Gulak, "Prospects for multiple-valued integrated circuits." *IEICE Trans. Electron.* E76-C(3), pp. 372–382, Mar. 1993.
2. Y. Yuminaka, T. Aoki, and T. Higuchi, "Design of set-valued logic networks for wave-parallel computing," in *Proc. 23rd Int. Symp. Multiple-Valued Logic*, May 1993, pp. 277–282.
3. Y. Yuminaka, T. Aoki, and T. Higuchi, "Design of wave-parallel computing architectures and its application to massively parallel image processing." *IEICE Trans. Electron.* E76-C(7), pp. 1133–1143, July 1993.
4. Y. Yuminaka, T. Aoki, and T. Higuchi, "Design of wave-parallel computing circuits for densely connected architectures," in *Proc. 24th Int. Symp. Multiple-Valued Logic*, May 1994, pp. 207–214.
5. M. P. Craven, K. M. Curtis, and B. R. Hayes-Gill, "Consideration of multiplexing in neural network hardware," in *IEE Proc.-Circuits Devices Syst.* 141(3), June 1994, pp. 237–240.
6. S. Sakurai, T. Aoki, and T. Higuchi, "Wire-free computing circuits using optical wave-casting," in *Proc. 25th IEEE Int. Symp. Multiple-Valued Logic*, May 1995, pp. 8–13.
7. T. Aoki, M. Kameyama, and T. Higuchi, "Interconnection-free biomolecular computing." *IEEE Computer* 25, pp. 41–50, Nov. 1992.
8. D. A. Simovici, I. Stojmenovic, and R. Tosic, "Functional completeness and weak completeness in set logic," in *Proc. 23rd Int. Symp. Multiple-Valued Logic*, May 1993, pp. 250–254.
9. K. Nakajima and Y. Hayakawa, "Correct reaction neural network." *Neural Networks* 6, pp. 217–222, June 1993.
10. H-J. Song, C-K. Kim, "An MOS four-quadrant analog multiplier using simple two-input squaring circuits with source followers." *IEEE J. Solid-State Circuits* SC-25, pp. 841–848, June 1990.
11. L. J. Pu and Y. P. Tsividis, "Transistor-only frequency-selective circuits." *IEEE J. Solid-State Circuits* SC-25, pp. 821–832, June 1990.

Yasuhi Yuminaka received the B.E., M.E., and D.E. degrees in electronic engineering from Tohoku University, Sendai, Japan in 1990, 1992, and 1995, respectively. He is currently a research associate in the Department of Electronic Engineering at Gunma University, Japan. His research interests include design of new paradigm computing system, especially multiple-valued logic system, and its application. Dr. Yuminaka received the IEE Ambrose Fleming Premium Award in 1994, the Niwa Memorial Award in 1995, and the Young Engineer Award from the Institute of Electronics, Information and Communication Engineers of Japan in 1995. He is a member of the IEEE.

Yoshisato Sasaki received the B.E., M.E., and D.E. degrees in electronic engineering from Tohoku University, Sendai, Japan, in 1962, 1964, and 1967, respectively. He is currently a Professor of the Department of Electronic Engineering at Gunma University. His general research interests include physics and technology of silicon processing. He is a member of the Japan Society of Applied Physics, the Physical Society of Japan.

Takafumi Aoki is an associate professor in the Graduate School of Information Sciences at Tohoku University, Japan. His research interests include computer arithmetic, VLSI computing structures for signal and image processing, multiple-valued logic systems, optical computing and molecular computing. Dr. Aoki received the Outstanding Paper Award at the 1989 IEEE International Symposium on Multiple-Valued Logic, the Outstanding Transactions Paper Award from the Institute of Electronics, Information and Communication Engineers of Japan in 1989, and the IEE Ambrose Fleming Premium in 1994. He received the B.E., M.E., and D.E. degrees in electronic engineering from Tohoku University in 1988, 1990, and 1992, respectively. He is a member of the IEEE Computer Society, the Society of Instrument and Control Engineers of Japan, and the Institute of Electronics, Information and Communication Engineers of Japan.

Tatsuo Higuchi received the B.E., M.E., and D.E. degrees in electronic engineering from Tohoku University, Sendai, Japan, in 1962, 1964 and 1969, respectively. He was a Professor in the Department of Electronic Engineering at Tohoku University, and is currently a Professor and Dean of the Graduate School of Information Sciences at Tohoku University. His general research interests include the design of 1-D and multi-D digital filters, linear time-varying system theory, fractals and chaos in digital signal processing, VLSI computing structures for signal and image processing, multiple-valued ICs, multiwave opto-electronic ICs, interconnection-free bio-electronic ICs and fault-tolerant computing. Dr. Higuchi received the Outstanding Paper Awards at the 1984, 1985, 1987, and 1989 IEEE International Symposiums on Multiple-Valued Logic, the Outstanding Transactions Paper Award from the Society of Instrument and Control Engineers of Japan in 1984, the Technically Excellent Award from the Society of Instrument and Control Engineers of Japan in 1986, the Outstanding Transactions Paper Award from the Institute of Electronics, Information and Communication Engineers of Japan in 1989, the Technically Excellent Award from the Robotics Society of Japan in 1990, and IEE Ambrose Fleming Premium Award in 1994. He is a member of the Society of Instrument and Control Engineers of Japan, and Robotics Society of Japan. He is an IEEE fellow.

www.ingramcontent.com/pod-product-compliance
Lightning Source LLC
LaVergne TN
LVHW062241200926
848484LV00006B/1584
9780792381259